Software Engineering Mathematics

Dedicated to Jamie, Monica, and Julia

Software Engineering Mathematics

Formal Methods Demystified

Jim Woodcock
Oxford University

Martin Loomes
Hatfield Polytechnic

Pitman Publishing
128 Long Acre, London WC2E 9AN

A Division of Longman Group UK Limited

First published in Great Britain 1988
Reprinted 1989, 1991

British Library Cataloguing in Publication Data

Woodcock, James
Software engineering mathematics.
1. Computer systems. Programs.
Mathematics
I. Title II. Loomes, Martin
005.13′1

ISBN 0-273-02673-9

Printed and bound in Great Britain by
Biddles Ltd, Guildford and King's Lynn

Contents

Preface

A few years ago, formal methods of software development were practised only by academics and consenting industrialists behind locked doors. Today things have changed: there is evidence that mathematically based techniques can be applied to real problems on an industrial scale. Indeed, there is a growing recognition that it is only by using mathematics that ways of subordinating complexity can be found.

Industry is now demanding people skilled in the application of the new methods such as VDM, Z, CCS, CSP, OBJ, and LOTOS. It is now presented with the problem that few software engineers have a firm grasp of the theoretical foundations underlying these techniques. Two things must be done to remedy this situation. Practising software engineers must be given the opportunity to understand the theoretical foundations and become proficient in applying the advanced methods, and universities and polytechnics must start to teach them.

This book makes the mathematical basis of formal methods accessible both to the student and to the professional. It is motivated in the later chapters by examples and exercises having a strong computing flavour, and by the conviction that mathematics is as essential to design and construction in software engineering as it is to other engineering disciplines. The exercises range from fairly simple drills, intended to provide familiarity with concepts and notation, to fairly demanding activities, which require a good grasp of the material.

The first four chapters of the book are devoted to foundations. We start with an introduction to formal systems, and then introduce the propositional and predicate calculi. We conclude the first part of the book with a chapter on theories in general. This section of the book is intentionally divorced from software engineering applications, as introducing problems before sufficient mathematical power has been developed might lend support to the apochryphal view that only trivial problems are amenable to formalisation.

The second part of the book builds upon the foundations by covering in detail theories of sets, relations, functions, and sequences. We introduce a mathematical rattle-bag of useful tools: mathematical data types and operations upon them. These mathematical data types are powerful enough to describe many aspects of software systems, and small case studies are included as examples of their use in the modelling of software: a configuration manager, a storage allocator, and a simple backing store interface. Rather than inventing yet another new language for the rattle-bag, we have chosen to use the concrete syntax of the Z notation[1]. The principles involved, however, are general ones, and the reader will have no difficulty in transferring them to other notations, such as VDM.

The third part of the book presents two detailed case studies in the use of mathematics in software engineering. The first is the specification of the behaviour of a telephone exchange, and the second illustrates the importance of the development of a mathematical theory in gaining an understanding of a system. Both case studies stress the rôles of modelling and of proof in the construction of specifications.

The final part of the book describes other techniques, showing that all that we have done is to use mathematics. The important thing is to choose the appropriate kind of mathematics and find a good style in its use. First we describe the algebraic approach, and then we summarise what formal methods are and give some examples, comparing and contrasting the different techniques.

Acknowledgments

We would like to thank all of our colleagues for providing the intellectual environment within which this book could be written. In particular, the members of the Z group at Oxford, including Steve King

[1] I. Hayes, *Specification Case Studies*, Prentice-Hall International, 1987.

(who also provided a thoroughly detailed review of the entire book), Carroll Morgan, Jeff Sanders, Ib Holm Sørensen, Mike Spivey (whose practical help with the book saved an immeasurable amount of time), and Bernard Sufrin; and the Software Engineering group at Hatfield, including Bob Dickerson, Richard Mitchell, and Wilf Nichols. Useful comments on an early draft were made by Dan Simpson. Many people have helped us to create this book; they all know who they are, and we should like to offer our sincere thanks. After a long and exhausting struggle, which we almost lost, the book was produced in camera-ready copy using LaTeX.

Chapter 1

Formal Systems

In this chapter we will introduce two important aspects of mathematics. First, mathematics can be viewed as a language for description and discussion. Second, mathematics embodies rules governing the manipulation of the abstract strings of symbols which constitute the descriptive language.

We will start by introducing the concept of a *formal system* which consists of a *formal language* and a *deductive apparatus*. This provides us with a way of generating and manipulating abstract strings of symbols. If the formal system is to be useful, however, we must give the strings in the formal language some meaning, or *semantics*, in the world in which we want to use it; this is achieved by providing an *interpretation* for the formal system.

You are asked to remember while reading this chapter that we are laying foundations for all that is to follow. Like most foundations, once they have been established they can be left buried beneath the surface, but, like most foundations, if they are badly laid then all that follows will eventually collapse around you!

1.1 Formal Languages

A formal language comprises two parts, its *alphabet* which specifies what symbols are to be found in the language, and its *syntax* which specifies how these symbols may be put together. An alphabet is specified by writing down the symbols in curly brackets, $\{\ldots\}$, separated by commas, with conventional shorthands such as $\{a, b, \ldots, z\}$ where there are many symbols involved. This small piece of notation is part of a *metalanguage* we are using for describing a formal language.

Example 1.1 *Here are the alphabets associated with some commonly used formal languages.*

- *A language for expressing real numbers, such as* " 210.356"
  $\{0, 1, 2, 3, 4, 5, 6, 7, 8, 9, .\}$
- *A language for expressing musical notes such as* "$B\flat$"
  $\{ A, B, C, D, E, F, G, \sharp, \flat, \natural\}$
- *A language for expressing playing cards such as* "$\clubsuit$10"
  $\{\clubsuit, \diamondsuit, \heartsuit, \spadesuit, A, 2, 3, 4, 5, 6, 7, 8, 9, 10, J, Q, K\}$
- *A language for expressing book section numbers such as* " 3.4.3"
  $\{0, 1, 2, 3, 4, 5, 6, 7, 8, 9, .\}$

When the alphabet of a formal language is obvious, we often take it for granted. It is usually only necessary to make it explicit where confusion may arise, such as when a symbol may be made up from from two or more parts.

We can see from these examples that sometimes two languages have the same alphabet. What makes them different languages, however, is that the grammar allows different strings to be formed from the alphabet.

Example 1.2 *Here are some examples of strings formed from the common alphabet used for real numbers and for section numbers in books.*

- *acceptable section numbers* " 3.2" " 1.2.3" " 1" "3.6.1"
- *not acceptable section numbers* " .7" "3."
- *acceptable real numbers* "23.7" "12" "12." ".76" ".7"
- *not acceptable real numbers* "3.6.1"

An acceptable string of symbols in a language is called a *well formed formula* or *wff* (pronounced 'woof') of that language. The formal language itself is just the collection of all of its well formed formulae. Exactly what constitutes a wff is defined by the *grammar rules* for the language, and these specify the *grammar* or *syntax* of the language.

To give a syntax we need a metalanguage for expressing grammar rules. This could be a natural language (that is, a language such as French or English) or another formal language.

Example 1.3 *The Language $\mathcal{L}$ has as its alphabet $\{\star, \diamond\}$, and its syntax (expressed in a metalanguage of English) is given by the grammar rule*

> *"A wff in $\mathcal{L}$ is any finite string of zero or more $\star$ symbols, followed by between one and four $\diamond$ symbols, or a string of one or more $\star$ symbols with no $\diamond$ symbols following."*

The following are examples of wffs in $\mathcal{L}$

$\star\star\star\star\star\star\star\star\star\star\diamond\diamond\diamond$

$\star\star\star\star\diamond\diamond\diamond\diamond$

$\diamond\diamond$

$\star\star\star\star\star\star\star\star\star\star\star\star\star\star\star\star\diamond$

$\star\star\star\star\star\star$

The following are not wffs in $\mathcal{L}$

$\star\star\star\diamond\diamond\diamond\star$

$\diamond\diamond\diamond\star$

It is important to realise that you should not be asking what the above wffs mean, since we have not yet discussed how to assign meanings to wffs in formal systems by giving a semantics. The relevant question to ask is whether a given string of symbols is a wff of the given language. If we are using English as our metalanguage then even this may not be very easy to answer, because natural language descriptions of things are often very hard to reason about accurately.

Exercise 1.1 *Study carefully the grammar rule given in Example 1.3, and see how many different ways of interpreting it you can find.*

To introduce more clarity and precision, we can choose another formal language as a metalanguage for specifying our grammar. The formal metalanguage we will use in this book is a subset of the *standard syntactic metalanguage*[1].

The grammar for a language is given by a number of rules, each of which defines and names a structural entity within the language. The naming is done solely to enable entities to be referred to in later rules, and the names are not part of the language being defined. The definition of the entity is done by following the name with "=" and then giving a rule defining its grammatical structure, terminated with ";". Whenever a symbol from the alphabet of the language appears, it is enclosed in quotation marks thus "...". When one of these entities can be made up in several different ways the choices are divided by the symbol "|".

Example 1.4 *The following rules create an entity called digit which can be any one of the symbols* 0...9, *and then use this to describe the entity hexadecimal digit.*

$$\begin{aligned} digit = {} & \texttt{"0"} \mid \texttt{"1"} \mid \texttt{"2"} \mid \texttt{"3"} \mid \texttt{"4"} \mid \texttt{"5"} \\ & \mid \texttt{"6"} \mid \texttt{"7"} \mid \texttt{"8"} \mid \texttt{"9"}; \end{aligned}$$

$$\begin{aligned} hexadecimal\ digit = {} & digit \mid \texttt{"}A\texttt{"} \mid \texttt{"}B\texttt{"} \mid \texttt{"}C\texttt{"} \\ & \mid \texttt{"}D\texttt{"} \mid \texttt{"}E\texttt{"} \mid \texttt{"}F\texttt{"}; \end{aligned}$$

In order to specify that items appear in some particular order we list them, separated by commas.

Example 1.5 *The following rule describes any of the strings* 00, 01, *up to* 99, *in terms of digit defined above*

$$two\ digit\ number = digit, digit;$$

---

[1] "An introduction and handbook for the standard syntactic metalanguage", National Physical Laboratory Report DITC February 1983.

Example 1.6 *Here is a language for writing down unsigned decimal numbers. Its alphabet is* $\{0,1,2,3,4,5,6,7,8,9,.\}$ *and its grammar rules are*

$$
\begin{aligned}
\textit{unsigned decimal} &= \textit{unsigned integer} \mid \textit{decimal fraction} \\
&\quad \mid \textit{unsigned integer}, \textit{decimal fraction}; \\
\textit{decimal fraction} &= \texttt{"."}, \textit{unsigned integer}; \\
\textit{unsigned integer} &= \textit{digit} \mid \textit{digit}, \textit{unsigned integer}; \\
\textit{digit} &= \texttt{"0"} \mid \texttt{"1"} \mid \texttt{"2"} \mid \texttt{"3"} \mid \texttt{"4"} \mid \texttt{"5"} \\
&\quad \mid \texttt{"6"} \mid \texttt{"7"} \mid \texttt{"8"} \mid \texttt{"9"};
\end{aligned}
$$

It is important to note that because we are *defining* the syntax for a language we cannot sensibly ask whether the grammar is right or wrong. All we can usefully ask is whether it captures the language we intended. For example, the definition of *unsigned decimal* above does not allow the string "23." as a wff, and we should ensure that it was our intention to exclude strings of this form from the language.

Exercise 1.2 *Use the standard syntactic metalanguage to write a definition of the syntax of a language for expressing unsigned decimals which will allow "23."as a wff, but will not allow ".34".*

Exercise 1.3 *Using the alphabet* $\{\clubsuit, \diamondsuit, \heartsuit, \spadesuit, A, 2, 3, \ldots, 10, J, Q, K\}$ *specify, in the standard syntactic metalanguage, a grammar defining a language for naming playing cards, such as "♣10"*

Exercise 1.4 *Use the standard syntactic metalanguage to write a definition of the syntax of a language for expressing hexadecimal numbers in the range* 00 *to* $7F$.

Exercise 1.5 *Use the standard syntactic metalanguage to give the grammar rules for the language* $\mathcal{L}$, *informally described in Example 1.3.*

Exercise 1.6 *Use the standard syntactic metalanguage to give the grammar rules defining a language suitable for expressing book section numbers.*

Exercise 1.7 *Given a language with alphabet* $\{\lambda, \bullet, (,), x, y, z\}$ *and syntax*

$$\begin{array}{lcl} expression & = & variable\ name \mid expression, expression \\ & & \mid "\lambda", variable\ name, "\bullet", expression \\ & & \mid "(", expression, ")"; \\ variable\ name & = & "x" \mid "y" \mid "z"; \end{array}$$

which of the following are expressions in the language?

1. $\lambda x \bullet yz$
2. $\lambda \bullet x \lambda \bullet y$
3. $\lambda y \bullet x \bullet z$
4. $\lambda x \bullet x(yz)$
5. $\lambda x \bullet \lambda y \bullet xyz$

1.2 Semantics

So far we have looked at ways of specifying meaningless strings of symbols that we wish to consider as a language. It is important to understand this, even if you are singularly unimpressed by the feat! You should note, for example, that the symbol "9" in the previous examples does not mean the number nine, in fact it does not mean anything at all yet. This might be easier to appreciate for the musical example above, since not everyone will understand what $A\flat$ might mean, although we are prepared to accept that it might mean something to a musician. Similarly the *meaning* of the "$\clubsuit A$" is not really known until we describe a particular game; it might be the lowest card in the suit or the highest.

The next stage is to give the formal language some *semantics*, by specifying the meaning of each wff admitted by the grammar. This is achieved by giving the language an *interpretation* which assigns a value from some domain of interest to each wff; we will say that this value is the meaning of the wff under the particular interpretation.

Exactly how we should write down interpretations is a complex question, and a discussion of this is beyond the scope of this book, so we will be content with a natural language description of what our wffs mean. This is far from perfect, and a formal metalanguage for the presentation of interpretations might be better, but until we have presented the rest of the book we do not have sufficient tools at our disposal to discuss suitable formal systems for this task.

Example 1.7 *Here is an interpretation, I_1, for the language $\mathcal{L}$ introduced in Example 1.3. Let*

$\star$ *denote* 5

$\diamond$ *denote* 1

and the placing of one symbol next to another indicate that their values under the interpretation are to be added together. Thus, for example,

$\diamond\diamond\diamond$ *denotes* $1 + 1 + 1$ *or* 3

$\star\star\star\diamond\diamond$ *denotes* $5 + 5 + 5 + 1 + 1$ *or* 17

So the interpretation I_1 of the language $\mathcal{L}$ assigns to each wff of $\mathcal{L}$ a value from the domain $1, 2, 3, \ldots$ As a shorthand for "in the interpretation I_1 $\diamond\diamond\diamond$ denotes 3" we will write $I_1(\diamond\diamond\diamond) = 3$.

This is only one interpretation of the language, here is another:

Example 1.8 *Here is an alternative interpretation, I_2, for the language $\mathcal{L}$. Let*

$I_2(\star) = 10$

$I_2(\diamond) = 2$

and the placing of one symbol next to another indicate that their values under the interpretation are to be added together, which is to say that $I_2(xy) = I_2(x) + I_2(y)$. Now

$I_2(\diamond\diamond\diamond) = 2 + 2 + 2 = 6$

$I_2(\star\star\star\diamond\diamond) = 10 + 10 + 10 + 2 + 2 = 34$

This interpretation assigns to each wff of $\mathcal{L}$ a value from the domain $2, 4, 6, 8, \ldots$

This ability to use pieces of formal notation to denote different things under different interpretations is what makes mathematics so powerful. The natural numbers $0, 1, 2, \ldots$, for example, if treated as wffs in a formal language, have many different interpretations. We can use "3" to stand for the number of apples in a dish, the number of cars in a car park, and so on. We could, if we wanted to be perverse, use it to stand for the number of hours in a day, but then we would be giving it an interpretation alien to most other users of the language. We will avoid choosing interpretations which are deliberately at odds with common usage.

1.3 Inference Systems

We have now reached the stage where we can define a formal language in terms of meaningless strings of symbols from its alphabet, and then assign meanings to these strings by giving an interpretation for the language. This allows us to describe things using a formal notation. We are now going to add to this descriptive power the facility to manipulate these formally defined strings of symbols by analysis of their syntactic shape, and without needing to know what they represent. This is achieved by adding a *deductive apparatus* to the formal language, and the result is called a *formal system.*

The most important requirement we shall make regarding this deductive apparatus is that it makes *no reference to any particular interpretation* of the formal language, but is concerned with wffs as meaningless syntactic objects. In this way we can carry out symbolic manipulations without needing to consider what our symbols stand for, just as in arithmetic we can add two numbers without worrying about wondering whether they denote numbers of apples or numbers of cars. The two components of a deductive apparatus are

- Axioms—wffs which can be written down without reference to any other wffs in the language.

- Inference rules—rules which allow us to produce wffs in the language as *immediate consequences* of other wffs.

This may sound very abstract, but abstraction is precisely what we are trying to achieve. Note that we do not say that axioms are true, since axioms are just wffs, and to say one denotes something that is true is to attribute meaning, which we are trying to avoid doing. Later we will see that we are usually only interested in those interpretations where axioms do denote true statements. We cannot, however, guarantee that all the users of our formal system will adhere to the rules; mathematics, like most useful tools, cannot be held responsible for the competence of its users.

Example 1.9 *Here is a simple formal system. The alphabet of the language is* $\{\star, \diamond, \circ\}$ *and its syntax is given by the grammar rules*

$$\begin{aligned} \textit{sentence} \quad &= \textit{string of stars}, \text{"} \diamond \text{"}, \textit{string of stars}, \\ &\quad \text{"} \circ \text{"}, \textit{string of stars}; \\ \textit{string of stars} &= \text{"} \star \text{"} \mid \textit{string of stars}, \text{"} \star \text{"} ; \end{aligned}$$

The deductive apparatus consists of

Axiom $\star \diamond \star \circ \star \star$

Rule *"If* $a \diamond b \circ c$ *is a given wff, where* a, b *and* c *are strings of stars, then* $a \diamond b \star \circ c \star$ *is an immediate consequence of it."*

Notice that we have resorted to English as a metalanguage for expressing our deductive apparatus, allowing ourselves the luxury of referring to syntactic components, such as *string of stars*, to make life easier. In later chapters we shall introduce a more formal metalanguage for this purpose.

The use of the rule in the deductive apparatus relies on matching patterns in wffs to identify a, b and c. This is no different from identifying the side opposite the right-angle before applying Pythagoras' Theorem in geometry.

Example 1.10 *To show that* $\star \diamond \star \star \star \star \circ \star \star \star \star \star$ *is an immediate consequence of the wff* $\star \diamond \star \star \star \circ \star \star \star \star$, *in the formal system of Example 1.9, we need to identify* a, b *and* c *in the rule.*

$$\overbrace{\star}^{a} \diamond \overbrace{\star \star \star}^{b} \circ \overbrace{\star \star \star \star}^{c}$$

so that, applying the rule we get

$$\overbrace{\star}^{a}\ \diamond\overbrace{\star\star\star\star}^{b}\circ\overbrace{\star\star\star\star}^{c}\ \star$$

We can now look at an interpretation of this formal system.

Example 1.11 *Here is an interpretation for the formal system above,*

Let $\star$ *denote* 1, $\star\star$ *denote* 2, $\star\star\star$ *denote* 3 ...

let $\diamond$ *denote the normal addition symbol* $+$

let $\circ$ *denote* $=$ *in arithmetic.*

so that sentences are interpreted as arithmetic expressions of the form

$$a+b=c$$

which can be either true or false. Then the axiom becomes

$$1+1=2$$

and the inference rule can be thought of as saying that

if $a+b=c$ *then* $a+(b+1)=(c+1)$.

Note that this is a sensible interpretation, since the axiom denotes something that is true and the rule will only ever generate more true wffs. This is not a property of the formal system alone, but of how it is being interpreted; changing the meaning of $\diamond$ so that it denotes "–", for example, would certainly not fit in with our intuitions regarding arithmetic.

1.4 Proofs and Theorems

A *proof* in a formal system, $\mathcal{F}$, is a finite sequence of wffs in the associated formal language, each of which is either an axiom of $\mathcal{F}$ or an immediate consequence of one or more preceding wffs (as determined by the inference rules of the system). A wff which can be proved within the formal system $\mathcal{F}$ is called a *theorem* of $\mathcal{F}$; all axioms of $\mathcal{F}$ are also theorems of $\mathcal{F}$. In this book we will introduce the notion of a *good proof* as one where a reader can see the justification for writing down each wff; this means that some commentary is required to explain each step.

Here is a theorem, and its proof, of the formal system introduced in Example 1.9. Each line of our proof will contain three parts, a line-number to enable reference to the line, the wff we want to write down, and a justification for being allowed to write it down, with references to previous lines in the proof where appropriate.

Theorem 1.1 *The theorem is*

$$\star\diamond\star\star\star\star\circ\star\star\star\star\star$$

Proof

1	$\star\diamond\star\circ\star\star$	Axiom
2	$\star\diamond\star\star\circ\star\star\star$	Inference rule applied to 1
3	$\star\diamond\star\star\star\circ\star\star\star\star$	Inference rule applied to 2
4	$\star\diamond\star\star\star\star\circ\star\star\star\star\star$	Inference rule applied to 3

QED

Notice that in proving the required result we have found other theorems on the way, namely $\star\diamond\star\star\circ\star\star\star$ and $\star\diamond\star\star\star\circ\star\star\star\star$.

In the above example we did not attempt to use any ideas of what the wffs might stand for; we worked entirely at the syntactic level. Once we have found a theorem, however, it might be sensible to ask what the theorem would mean in some particular interpretation of the formal system.

Example 1.12 *Using the interpretation of Example 1.11, the above theorem becomes*

$$1 + 4 = 5$$

which seems consistent with our interpretation of the axiom and inference rule.

In fact, any theorem we can prove using this formal system with this interpretation will correspond to a true statement, and so we say that this formal system is *consistent* with its interpretation. Not all interpretations will have this property. The consistency of a formal system with its interpretation is an example of a *metatheorem*, as it states something *about* the system rather than being just a string of symbols *within* it. In the rest of this book we will only be interested in consistent interpretations of formal systems.

Now let us try something more ambitious, is $\star\star\star\diamond\star\star\star\star\circ\star\star\star\star\star\star\star$ a theorem? Since we know it can have an interpretation of

$$3 + 4 = 7$$

it might seem likely.

Alas, we hit the problem of our formal system being rather too simple to handle expressions with anything other than a single $\star$ before the $\diamond$. Even though $\star\star\star\diamond\star\star\star\star\circ\star\star\star\star\star\star\star$ is a wff of the language, and we can give it an interpretation, we cannot use our deductive apparatus to generate it as a theorem. This is because our formal system is *incomplete*. If it were complete we could prove as theorems all results we think of as true in the interpretation.

An interpretation of a formal system in which wffs denote statements which can be true or false is

- *consistent* if every theorem of the system interprets to a true statement

- *complete* if every true statement can be proved as a theorem.

Unfortunately most useful formal systems are incomplete, so there will be occasions when we will be unable to prove things we know to be true.

Perhaps some of you are forming the opinion that mathematicians spend a large proportion of their lives dreaming up formal systems, on the off-chance that someone will find useful interpretations for them. In general, of course, this is not the case; formal systems are usually designed with interpretations in mind.

Exercise 1.8 *Design a formal system in which you can prove statements such as* $3 + 4 = 7$ *to be theorems.*

1.5 Derivations

A *derivation* of the wff $\mathcal{W}$ in formal system $\mathcal{F}$ from a given set $\mathcal{P}$ of wffs, called *premises*, is a finite sequence of wffs from the language of $\mathcal{F}$ such that the last wff is $\mathcal{W}$ and each wff in the sequence is one of the following

- an axiom of $\mathcal{F}$
- a premise, that is, a wff from the given set $\mathcal{P}$
- an immediate consequence of one or more wffs preceding it in the sequence, as determined by the inference rules of $\mathcal{F}$.

This can be thought of as corresponding to an argument of the form "if we are given that ... then ...". If there is such a derivation of $\mathcal{W}$ from $\mathcal{P}$, we write $\mathcal{P} \vdash \mathcal{W}$. The symbol $\vdash$ is called the *syntactic turnstile*, and is a metasymbol in the sense that it is not part of the formal language itself, but allows us to say things *about* the formal system. The format we will use for derivations is identical to that for proofs, except that we will allow premises to be written down as well as axioms.

Derivation 1.1 *Using the formal system of Example 1.9:*

★★★◇★★○★★★★★ ⊢ ★★★◇★★★★○★★★★★★★

Derivation

1	★★★◇★★○★★★★★	premise
2	★★★◇★★★○★★★★★★	1 Rule 1
3	★★★◇★★★★○★★★★★★★	2 Rule 1

QED

Exercise 1.9 *Show that*

★◇★★○★★★★★ ⊢ ★◇★★★★○★★★★★★★

If we look at interpretations for the wffs in the above exercise, we seem to be saying that from "1+2=5" we can get "1+4=7". Since, in a sensible world, we will never start from such a false premise, the derivation of something equally false need not worry us. In particular, we must not be tempted to say there is something wrong with the formal system, since there may well be interpretations for which this derivation is very useful. We might be well advised to remember the old adage "poor workmen always blame their tools" when we incorrectly apply mathematics to problems.

Like many pieces of mathematics, it is sometimes convenient to reverse the way in which things are written down, and so $\mathcal{W} \dashv \mathcal{P}$ is exactly the same as $\mathcal{P} \vdash \mathcal{W}$, just as $7 > 3$ and $3 < 7$ are equivalent. In the special case when $y \vdash x$ and $x \vdash y$ (that is, x can be derived from y, and y can be derived from x) then we write $x \dashv\vdash y$.

Every proof in a formal system $\mathcal{F}$ is also a derivation from every set of premises $\mathcal{P}$, since there is no need in a derivation to make use of all, or indeed any, of the premises. We can view a proof as a special kind of derivation, one in which the set of premises is empty. In this case we use the syntactic turnstile to denote theorems, such as the one proved in Theorem 1.1 above, by $\vdash \star\diamond\star\star\star\star\circ\star\star\star\star\star$, remembering that theorems and derivations must always exist within a particular formal system, even if this is not explicitly identified.

It is important to understand the relationship between proofs and derivations . If $\mathcal{P} \vdash \mathcal{W}$ in some formal system $\mathcal{F}$, then $\vdash \mathcal{W}$ would hold in some formal system $\mathcal{F}'$, which contains all the axioms and inference rules of $\mathcal{F}$, together with the wffs $\mathcal{P}$ as additional axioms. Thus a derivation in one formal system corresponds to a proof in another, richer, system.

1.6 Summary

In this short chapter we have introduced the concept of a formal system. We have seen that a formal system comprises a formal language and a deductive apparatus, and that to be useful it must be given a semantics. The notions of proofs and derivations have been introduced, and a way of presenting these has been established. We have also introduced two properties of formal systems and their interpretations, completeness and consistency.

Chapter 2

Propositional Calculus

In this chapter we are going to investigate a formal system which has been developed for discussing the truth or falsity of a particular kind of statement called a *proposition*. We will start by explaining what propositions are, then we will introduce a formal language called *propositional logic* which has as the usual interpretation of its well formed formulae the truth values of propositions, and finally we will provide a deductive apparatus to create a formal system called *propositional calculus*.

2.1 Propositions

A proposition is a statement of some alleged fact which must be either true or false, and not both. This very simple sounding explanation, however, hides a number of very deep problems of philosophy and linguistics which we shall attempt to avoid in this book. For example, we will avoid expressions which involve reference points such as time and person, because statements like "my dog is a poodle" express different propositions depending on who says it, just as "it will rain tomorrow" depends on when it is said, and "this is a beautiful country" makes reference to a geographical location and a subjective view of beauty. Expressions such as these are not necessarily excluded from representing propositions, but they make the problem of discussing them much more difficult and will not be considered here.

To see if a piece of language represents a proposition we first check that it asserts some fact; thus questions and commands, for example, are not usually ways of expressing propositions. If some fact is being asserted then we must also check that the following two laws hold.

Law 2.1 (Excluded Middle) *A proposition is true or false, there can be no middle ground.*

Law 2.2 (Contradiction) *A proposition cannot be both true and false.*

Example 2.1 *The following are propositions*

- *elephants are mammals*
- *some birds can fly*
- *France is in Asia*

The following are not propositions

- *go away*
- *are you happy?*

The following lead to lengthy discussions, and will be ignored here!

- *wasps sting*
- *I think therefore I am*

It is important to realise that expressing a proposition requires the use of a language. Expressing the same proposition in several ways, however, perhaps using different languages, does not alter the fact that we still only have one proposition.

Example 2.2 *The following are all ways of expressing the same proposition*

- *five is bigger than four*
- *four is less than five*
- $5 > 4$
- *there is some positive number such that if we add it to four then the result would be five*

We are now going to introduce a formal language called *propositional logic*, which will allow us to discuss propositions while abstracting away from unnecessary details of representation. The well formed formulae of this language will be given interpretations of *truth values*, that is, whether the propositions they denote are true or false. This means that when we refer to the *proposition* P we do not mean the string of symbols used to express it but the truth value of the assertion that is being made. It is meaningless to ask how many words there are in P, or whether P is expressed in good English.

Sometimes a proposition can be expressed in a way which brings out the fact that its truth depends on the truth of other propositions. For example

John is cold and wet

could be re-phrased as

John is cold and John is wet

This now seems to assert two things, *John's coldness*, and *John's wetness*, rather than just his *cold-and-wetness*. This is a very subtle distinction and the way in which we choose to view such expressions is really up to us. If we decide that, for some particular task, it is useful to consider a proposition as consisting of two or more component propositions then it is said to be *compound*; a proposition which we have decided not to consider as compound is called *simple*. We are going to insist that compound propositions satisfy the following law.

Law 2.3 (Truth Functionality) *The truth value of a compound proposition is uniquely determined by the truth value of its constituent parts.*

Consider, for example, the statement

John is cold *because* he is wet.

The truth or falsity of the two propositions "John is cold" and "John is wet" tell us nothing about the truth of "John is cold *because* he is wet". Even if we know that John is both cold and wet, we know nothing about the *reason* for his coldness. He might have fallen into a lake of freezing water, so that his wetness did cause his coldness but he might be cold from the wind and be in a bath of hot water trying to warm up. Compound propositions which involve causality are not truth functional, and will not be considered further in this book.

2.2 Propositional Logic

Here is the formal language we are going to use to represent propositions. The alphabet is

$$\{P, Q, R, \ldots, P_1, P_2, \ldots, \wedge, \vee, \Rightarrow, \Leftrightarrow, \neg\}$$

where we include enough symbols from $P, Q, R, \ldots$, subscripted if necessary, to ensure that we never run out.
The syntax is given by the grammar rule

$$\begin{aligned} \mathit{sentence} = {} & \text{"}P\text{"} \mid \text{"}Q\text{"} \mid \text{"}R\text{"} \mid \ldots \mid \text{"}P_1\text{"} \mid \text{"}Q_1\text{"} \ldots \\ & \mid \text{"}\neg\text{"}, \mathit{sentence} \\ & \mid \text{"(", } \mathit{sentence}, \text{" } \vee \text{ ", } \mathit{sentence}, \text{")"} \\ & \mid \text{"(", } \mathit{sentence}, \text{" } \wedge \text{ ", } \mathit{sentence}, \text{")"} \\ & \mid \text{"(", } \mathit{sentence}, \text{" } \Rightarrow \text{ ", } \mathit{sentence}, \text{")"} \\ & \mid \text{"(", } \mathit{sentence}, \text{" } \Leftrightarrow \text{ ", } \mathit{sentence}, \text{")"}; \end{aligned}$$

Exercise 2.1 *Which of the following are sentences in propositional logic?*

1. $(P \Leftrightarrow Q)$
2. $P \Leftrightarrow Q$
3. $P \wedge\!\Rightarrow P_1$
4. $((P \wedge Q) \vee \neg R)$
5. $P \wedge Q \vee R \Leftrightarrow R$
6. $((P \Leftrightarrow Q) \Rightarrow ((P \wedge Q) \vee \neg R))$

You will notice that this formal language has a proliferation of brackets. We will allow ourselves the immediate luxury, at the expense of total formality, of dropping the outermost brackets which arise around whole sentences, thus we will write $P \wedge Q$ even though the grammar rule strictly says we should write $(P \wedge Q)$.

In the interpretation we are going to use for this language, the symbols $P, Q, R \ldots$ denote the truth values of simple propositions. The symbols $\wedge, \vee, \neg, \Rightarrow, \Leftrightarrow$ are interpreted as providing ways in which compound propositions can be built from simpler ones, and they are usually referred to as ***connectives***. The truth value of sentences of the form $(\mathcal{A})$, $\neg\mathcal{A}$, $(\mathcal{A} \wedge \mathcal{B})$, $(\mathcal{A} \vee \mathcal{B})$, $(\mathcal{A} \Rightarrow \mathcal{B})$ and $(\mathcal{A} \Leftrightarrow \mathcal{B})$, where $\mathcal{A}$ and $\mathcal{B}$ denote arbitrary sentences of the language, depend on the truth values of the component propositions. Since each compound form has only one or two constituent propositions, and we are insisting that the law of truth functionality be adhered to, it is a simple matter to relate the truth values of these forms to the truth values of their components, which will themselves be either simple or compound propositions. In theory, the connectives could be given different meanings in every interpretation, but in practice this does not happen, and only the meanings of $P, Q, R \ldots$ will vary.

A concise way of giving the meaning of compound forms is in tabular form. For those sentences which involve only one other proposition, the table

$\mathcal{A}$	$(\mathcal{A})$	$\neg\mathcal{A}$
T	T	F
F	F	T

shows that if $\mathcal{A}$ is interpreted as denoting some simple or compound proposition which is true, then $(\mathcal{A})$ is also true but $\neg\mathcal{A}$ is false, and if $\mathcal{A}$ is false then so is $(\mathcal{A})$ but $\neg\mathcal{A}$ is true.

Sentences containing the other connectives can be given interpretations in the same way, except that they involve two component propositions, which we will call $\mathcal{A}$ and $\mathcal{B}$.

$\mathcal{A}$	$\mathcal{B}$	$\mathcal{A} \wedge \mathcal{B}$	$\mathcal{A} \vee \mathcal{B}$	$\mathcal{A} \Rightarrow \mathcal{B}$	$\mathcal{A} \Leftrightarrow \mathcal{B}$
T	T	T	T	T	T
T	F	F	T	F	F
F	T	F	T	T	F
F	F	F	F	T	T

Thus assigning a truth value to a sentence containing connectives involves the following steps:

- giving an interpretation to the symbols denoting simple propositions
- using the above tables to evaluate a truth value for the sentence.

If the compound proposition contains more than one connective, then we interpret the brackets as implying an order of evaluation, and evaluate inside the brackets first[1]. In the absence of brackets indicating otherwise, the evaluation of negations are always carried out before any other connective: hence the reason our grammar rule does not insist on $(\neg \mathcal{A})$ but allows $\neg \mathcal{A}$.

Example 2.3 *If we interpret P as the truth value of some true proposition and Q as the truth value of some false proposition, then to evaluate $\neg((P \vee Q) \Rightarrow Q)$ we must first reveal its structure.*

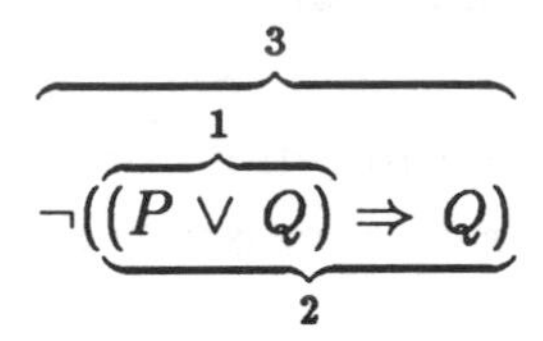

We now have to evaluate something of the form $(\mathcal{A} \vee \mathcal{B})$, where $\mathcal{A}$ denotes a true proposition and $\mathcal{B}$ denotes a false one. Using the above table we can see that this evaluates to true, so now we can evaluate the sentence of the form $(\mathcal{A} \Rightarrow \mathcal{B})$ where $\mathcal{A}$ denotes a true proposition and $\mathcal{B}$ denotes a false one; this evaluates to false. The outermost wff, $\neg \mathcal{A}$, where $\mathcal{A}$ denotes the truth value of a false proposition, evaluates to true. Thus,with the given interpretation of P and Q, we can determine that $\neg((P \vee Q) \Rightarrow Q)$ denotes the truth value of a true expression.

This way of laying out such a process is clearly not very satisfactory, so we will introduce the idea of a *truth table*. This allows us to tabulate the above process in an ordered way. It also allows us to discuss the interpretation for a complex sentence without actually deciding on the interpretations for its simple components, by considering all possible interpretations.

[1] An alternative approach here would be to define a set of *precedence rules* for the connectives, just as in the arithmetic expression $2 \times 3 + 4$ the $\times$ takes precedence over the $+$. In this case, the grammar could be modified to allow sentences without so many brackets.

To do this we draw up a table containing a row for each interpretation of the relevant simple propositional symbols, $P, Q, R \ldots$ contained in the sentence that we want to consider, and a column for every sub-expression used in building up the complex expression.

Example 2.4 *Here is a truth table showing all possible interpretations of the expression,* $\neg((P \vee Q) \Rightarrow Q)$. *It has four rows, corresponding to the four different combinations of truth values that P and Q can take, and columns corresponding to the formations of* $(P \vee Q)$, $((P \vee Q) \Rightarrow Q)$ *and* $\neg((P \vee Q) \Rightarrow Q)$

P	Q	$(P \vee Q)$	$((P \vee Q) \Rightarrow Q)$	$\neg((P \vee Q) \Rightarrow Q)$
T	T	T	T	F
T	F	T	F	T
F	T	T	T	F
F	F	F	T	F

It is important to note that the above truth table gives us *the four different interpretations* for the sentence $\neg((P \vee Q) \Rightarrow Q)$ corresponding to the four different possible combinations of P and Q. If we are only interested in certain interpretations then we do not need to produce all the rows in the table.

Exercise 2.2 *How many rows will there be in a truth table showing all possible interpretations of a sentence containing n simple propositions?*

Example 2.5 *Here is the truth table to show all possible interpretations of* $((P \wedge Q) \vee \neg R) \Leftrightarrow P$

P	Q	R	$(P \wedge Q)$	$\neg R$	$((P \wedge Q) \vee \neg R)$	$((P \wedge Q) \vee \neg R) \Leftrightarrow P$
T	T	T	T	F	T	T
T	T	F	T	T	T	T
T	F	T	F	F	F	F
T	F	F	F	T	T	T
F	T	T	F	F	F	T
F	T	F	F	T	T	F
F	F	T	F	F	F	T
F	F	F	F	T	T	F

Exercise 2.3 *Produce truth tables to show all possible interpretations of the following*

1. $P \wedge (P \vee Q)$

2. $(P \vee Q) \wedge (P \Rightarrow Q)$

3. $\neg P \wedge (P \vee (Q \Rightarrow P))$

4. $(P \wedge (Q \vee P)) \Leftrightarrow P$

5. $(P \Rightarrow Q) \Rightarrow (\neg P \vee Q)$

6. $((P \Rightarrow Q) \wedge (R \Rightarrow S) \wedge (P \vee R)) \Rightarrow (Q \vee S)$

We have now seen how to produce the truth table associated with a sentence which denotes a compound proposition, but how might such a proposition be expressed in English? One way to answer this question is to consider each connective in turn and see how it might arise.

The connective $\neg$ is usually referred to as *not* because the English expression of propositions containing it frequently use that word. It has the property that if some sentence $\mathcal{A}$ denotes the truth value of a proposition then $\neg\mathcal{A}$ denotes the truth value of the "opposite" proposition. For example, if we interpret the sentence P in our logic as denoting the proposition expressed by "all cats are mammals", then the sentence $\neg P$ denotes, in a compound form, the proposition expressed by "not all cats are mammals". We could equally well choose to view this as a simple proposition and denote it by Q, but then we would have chosen to ignore any connection between the two propositions. The sentence $\neg P$ is referred to as the *negation of P*.

The connective $\wedge$ is usually referred to as *and*, since it frequently arises when we denote a proposition in which we are asserting two things, such as "it is raining *and* John is wearing a hat". If we denote "it is raining" by P and "John is wearing a hat" by Q, then the compound proposition "it is raining and John is wearing a hat" can be denoted by $P \wedge Q$. The sentence $P \wedge Q$ is referred to as the *conjunction of* P and Q.

The connective $\vee$ is used to capture the idea of an *inclusive or*[2], that is to say a sentence $P \vee Q$ is interpreted as the truth value of the compound proposition "P or Q or both P and Q". This is rather difficult to illustrate by representations in English, since the word "or" is frequently used with a fairly loose interpretation. For example, "John drinks tea or John drinks coffee" might well mean John is indifferent (thus he drinks both tea and coffee) or it might imply that the speaker knows John only drinks one of these beverages but cannot remember which: we usually depend on context to differentiate between these uses. $P \vee Q$ is also referred to as the *disjunction* of P and Q.

It is also hard to give an English equivalent for the connective $\Rightarrow$. It is usually thought of as *implication* or *if ... then ...*, but both of these are inadequate to capture the subtleties of sentences which contain this connective. It is better to think of its interpretation only in terms of the tabular definition given above. From this we can see that a sentence of the form $\mathcal{A} \Rightarrow \mathcal{B}$ is true if both $\mathcal{A}$ and $\mathcal{B}$ are true, and also if $\mathcal{A}$ is false, regardless of the truth of $\mathcal{B}$. Thus a sentence capturing

mice are fish $\Rightarrow$ Mars is made of green cheese

will have an interpretation of true since mice are not fish! We would once again remind the reader that this is not an error in the formal system, but an example of a careless use of it. If confronted with a form like this we should ask exactly why we wrote it down, and not why the mathematics looks strange.

Thinking of $\Rightarrow$ as an "if $\cdots$ then $\cdots$" construct tends to lead to its incorrect use by assuming an unstated continuation of "otherwise $\cdots$". For example, a father might say to his child "if you do that again then I will take the toy away" which carries an implicit message that otherwise the child can retain the toy. The formalism is not so benign, however, and contains no information as to the father's behaviour in this case.

Many of these confusions arise because of a mistaken belief that we are able to capture causality, but as we have already seen this

[2] Our language does not have a connective symbol for exclusive or, but we can still capture the concept, "P or Q but not both", as $(P \vee Q) \wedge \neg(P \wedge Q)$.

would lead us to compound forms which are not truth functional, and hence are not admitted within propositional logic.

The connective $\Leftrightarrow$ is rather easier to understand. The tabular definition of it shows that it is true precisely when the things it connects are equal in the sense that they have the same truth values. The only warning here is to beware of thinking of $\Leftrightarrow$ as a way of writing "=", there are important differences which will be introduced later. At present, it is sufficient to note that "=" is not even in the alphabet of our formal language, so we must not even think of writing it down. This connective is sometimes called *double implication.*

2.3 Classifying Sentences

Some sentences, *because of their form*, will always interpret to true. For example, any sentence of the form "$\mathcal{A} \vee \neg\mathcal{A}$" will always be true as it is effectively just a statement of the law of the excluded middle. A sentence which takes the value true for every interpretation of its constituent parts is said to be *valid* or is called a *tautology.*

Similarly some sentences will always interpret to false, such as any of the form "$\mathcal{A} \wedge \neg\mathcal{A}$": if this were not the case then $\mathcal{A}$ could not be a proposition as it would break the law of contradiction. A sentence which always interprets to false, regardless of the truth values of its constituent parts, is called an *inconsistency* , or is said to be *inconsistent.*

A sentence which is neither a tautology nor a contradiction, that is to say it will take the values true and false under different interpretations, is said to be a *contingency.* The truth table drawn up for $\neg((S \vee T) \Rightarrow T)$ above shows this sentence to be a contingency; it is contingent on the values of S and T.

Sentences which take the value true under at least one possible interpretation are said to be *consistent.* Thus tautologies and contingencies are consistent, but inconsistencies are not.

Exercise 2.4 *Classify the sentences of Exercise 2.3 under the headings valid, consistent, contingent, and inconsistent.*

2.4 The Semantic Turnstile

If, for some list of sentences $\mathcal{P}$, whenever all the sentences of $\mathcal{P}$ are true then the sentence $\mathcal{W}$ is true, then we say that $\mathcal{W}$ is a *semantic consequence* of the set of sentences $\mathcal{P}$. We write this as $\mathcal{P} \models \mathcal{W}$. The metasymbol $\models$ is called the *semantic turnstile*.

Example 2.6 *Here is a truth table which shows that*

$$P, P \Rightarrow Q \models Q$$

P	Q	$P \Rightarrow Q$
T	T	T
T	F	F
F	T	T
F	F	T

Thus $P, P \Rightarrow Q \models Q$ since whenever the two sentences on the left hand side are interpreted as true (namely in row one) the sentence on the right is true.

In the special case when the set $\mathcal{P}$ is empty we read $\models \mathcal{W}$ as meaning that $\mathcal{W}$ is true for all interpretations. Thus a short-hand notation for "$\mathcal{W}$ is a tautology" is to write $\models \mathcal{W}$. This is reflected in a truth table by the column representing $\mathcal{W}$ containing the value true for all interpretations.

Exercise 2.5 *Draw truth tables to show the following*

1. $\neg P, P \vee Q \models Q$
2. $P \wedge Q \models P \vee Q$
3. $\models (P \Rightarrow Q) \Leftrightarrow (\neg P \vee Q)$
4. $((Q \vee P) \Rightarrow R) \models Q \Rightarrow R$
5. $\models P \Rightarrow (P \vee Q)$

It is important to realise that we are currently working at the semantic level, since we are using interpretations as captured by truth tables. This should be compared with our use of the syntactic turnstile earlier where we were working purely with strings of symbols, without any reference to their possible meanings.

2.5 Equivalence

Two sentences are said to be *equivalent* if and only if their truth values are the same under *every* interpretation. If $\mathcal{A}$ is equivalent to $\mathcal{B}$ we will introduce another metasymbol, and write $\mathcal{A} \equiv \mathcal{B}$.

Example 2.7 *Here we show that* $(P \Rightarrow Q) \equiv \neg(P \wedge \neg Q)$*, by drawing up the truth table*

P	Q	$P \Rightarrow Q$	$\neg Q$	$P \wedge \neg Q$	$\neg(P \wedge \neg Q)$
T	T	T	F	F	T
T	F	F	T	T	F
F	T	T	F	F	T
F	F	T	T	F	T

The equivalence can be seen from this, since the columns for $P \Rightarrow Q$ *and* $\neg(P \wedge \neg Q)$ *are identical.*

This piece of notation is by no means universal, and many writers on logic do not introduce the idea of equivalence at all. This is because of a property regarding equivalence in this formal system, that $\mathcal{A} \equiv \mathcal{B}$ is the case precisely when $\models \mathcal{A} \Leftrightarrow \mathcal{B}$.

Exercise 2.6 *Convince yourself of this fact by adding to the truth table of Example 2.7 a column for* $(P \Rightarrow Q) \Leftrightarrow \neg(P \wedge \neg Q)$.

For this reason many users of logic slip into the habit of using the same symbol both where we use "$\Leftrightarrow$" and to denote equivalence. This overloads the symbol, however, using it within the alphabet of the formal system and also as a metasymbol to discuss the interpretation of sentences. We prefer to keep the two concepts clearly distinct at this stage. In particular, the sentence $\mathcal{A} \Leftrightarrow \mathcal{B}$ is written down in the full knowledge that it may denote true or false in some interpretation, whereas $\mathcal{A} \equiv \mathcal{B}$ is an expression of a fact which the writer thinks is true.

There are a number of common properties of this interpretation of propositional logic that can be expressed as equivalences, and some of these are given in Figure 2.1(page 42).

Exercise 2.7 *Show the following equivalences by drawing up truth tables*

1. $P \wedge P \equiv P$

2. $P \Rightarrow Q \equiv Q \vee \neg P$

3. $(P \Rightarrow Q) \Rightarrow P \equiv P$

Before leaving the subject of interpreting sentences of propositional logic, we must stress that this is only one interpretation for one formal language. There are many slightly different languages which are frequently called propositional logic, because they too provide a language for discussing propositions. They may have different symbols denoting the connectives, such as & instead of $\wedge$, or a different set of connectives containing, for example, an exclusive or. They may have precedence rules in place of brackets. In theory, the interpretations of the connectives could be different, allowing $\Rightarrow$ to denote conjunction, for example, but this would be a very unconventional choice, and not very useful or productive.

2.6 Propositional Calculus

The preceding section has presented a language, together with its conventional interpretation, which enables us to discuss the truth values of propositions. Any reasoning that has taken place, however, has hinged on truth tables, and hence has occurred after interpretations have been made. In fact, of course, truth tables allow us to consider all possible interpretations for propositional symbols, so the discussion is still fairly general in nature. For sentences containing a large number of different propositional symbols, however, the tables become unmanageable. If the language is to provide the basis of a general method for discussing propositions then we need to provide a deductive apparatus, which enables reasoning to take place at the purely syntactic level, before interpretations are considered, rather than having to consider all possible interpretations explicitly.

The addition of a deductive apparatus to a propositional logic gives us a formal system which we will refer to as a *propositional calculus*. Just as there are many choices of language suitable for representing propositions, so too there are many choices of deductive apparatus for a propositional calculus. The one we are going to use is called a *natural deduction* system[3].

The rules are presented below in a very simple metalanguage. A rule takes the form of a list of sentences above a horizontal line, with one sentence under the line. This construct means that if we already have all the sentences above the line then we are allowed to derive, as an immediate consequence, the one below, citing the appropriate rule name for explanation.

The system is based on the idea of providing rules of inference for introducing and removing each of the five connective symbols. There are no axioms in this system, but the derivation rules are phrased in such a way that we can still achieve proofs of theorems, although we will need to modify our notion of a proof slightly. We will present the rules in two batches, first those where it is quite easy to see what they will reflect in an interpretation, and so are fairly intuitive, and then those which are more difficult to understand. In these rules $\mathcal{A}$ and $\mathcal{B}$ denote any simple or compound propositions.

∧–INTRODUCTION

$$\frac{\mathcal{A},\mathcal{B}}{\mathcal{A}\wedge\mathcal{B}} \quad \text{and} \quad \frac{\mathcal{A},\mathcal{B}}{\mathcal{B}\wedge\mathcal{A}}$$

∧–ELIMINATION

$$\frac{\mathcal{A}\wedge\mathcal{B}}{\mathcal{A}} \quad \text{and} \quad \frac{\mathcal{A}\wedge\mathcal{B}}{\mathcal{B}}$$

∨–INTRODUCTION

$$\frac{\mathcal{A}}{\mathcal{A}\vee\mathcal{B}} \quad \text{and} \quad \frac{\mathcal{A}}{\mathcal{B}\vee\mathcal{A}}$$

¬–ELIMINATION

$$\frac{\neg\neg\mathcal{A}}{\mathcal{A}}$$

[3] This system of logic is sometimes called a *Gentzen Natural Deduction* system after the logician who first suggested its use.

$\Rightarrow$–ELIMINATION

$$\frac{\mathcal{A}, \mathcal{A} \Rightarrow \mathcal{B}}{\mathcal{B}}$$

$\Leftrightarrow$–ELIMINATION

$$\frac{\mathcal{A} \Leftrightarrow \mathcal{B}}{\mathcal{A} \Rightarrow \mathcal{B}} \quad \text{and} \quad \frac{\mathcal{A} \Leftrightarrow \mathcal{B}}{\mathcal{B} \Rightarrow \mathcal{A}}$$

$\Leftrightarrow$–INTRODUCTION

$$\frac{\mathcal{A} \Rightarrow \mathcal{B}, \mathcal{B} \Rightarrow \mathcal{A}}{\mathcal{A} \Leftrightarrow \mathcal{B}}$$

We will now use this subset of the formal system to perform some derivations. We are working at a purely syntactic level, so that to identify and use an applicable rule we need to find a rule that matches, in grammatical structure, the situation we are trying to handle.

Derivation 2.1 *Show that*

$$P \wedge Q \vdash P \vee Q$$

Derivation

1	$P \wedge Q$	premise
2	P	1 by $\wedge$–elimination
3	$P \vee Q$	2 by $\vee$–introduction

QED

In this example the strategy for the derivation was easy to find. The premise is of the form $\mathcal{A} \wedge \mathcal{B}$, so an obvious first step is to eliminate the connective as it is not required on the right hand side. To introduce the $\vee$ symbol it is just a matter of using the appropriate introduction rule. Most derivations are not as easy as this, and there will be a number of possible things that might be tried at every stage. This means that an inexperienced user of propositional calculus will probably make many attempts before a derivation comes out 'correctly'. The formal system is not very complex, but finding an appropriate proof strategy is seldom a trivial problem, and can only be learned by experience. *Do not be disheartened by the seeming ease with which books present proofs; the authors of these proofs have without doubt spent many hours, and covered many pages, in the pursuit of derivations which are then tidied and presented in a clear and concise form.*

Derivation 2.2 *Show that*

$$P\,,\,Q\,,\,(P \wedge Q) \Rightarrow R\ \vdash R$$

Derivation

1	P	premise
2	Q	premise
3	$(P \wedge Q) \Rightarrow R$	premise
4	$P \wedge Q$	1,2 by $\wedge$–introduction
5	R	3,4 by $\Rightarrow$–elimination

QED

Derivation 2.3 *Show that*

$$P\,,\,P \Leftrightarrow Q \vdash Q$$

Derivation

1	P	premise
2	$P \Leftrightarrow Q$	premise
3	$P \Rightarrow Q$	2 by $\Leftrightarrow$–elimination
4	Q	1,3 by $\Rightarrow$–elimination

QED

Exercise 2.8 *Perform the following derivations.*

1. $(P \wedge Q) \wedge R \vdash P \wedge (Q \wedge R)$

2. $P\,,\,P \Rightarrow Q\,,\,Q \Leftrightarrow R \vdash Q \wedge R$

3. $\neg\neg Q \vdash Q \vee R$

4. $P \wedge Q\,,\,P \Rightarrow S\,,\,Q \Rightarrow T \vdash S \wedge T$

Some of the rules make use of ***assumptions***. These are internal to the derivation or proof, and must not be confused with premises. An assumption is only to be used for the purpose of deriving the particular result determined by the rule which motivated its introduction. Each assumption has a ***scope*** which runs from the line in which it is made to the line before the rule in which it is discharged: we will show this by ruling a line alongside every portion of a proof or derivation that is in the scope of an assumption. Assumptions must on no account be used outside their scope. Our definitions of "derivation", and "proof", now need to be modified slightly in order to allow us to proceed from assumptions. In addition to allowing axioms and premises to be written down we will also allow any assumptions to be made but noting that anything written down within the scope of an assumption is not to be taken as derived or proved.

In the rules that follow we will use $\mathcal{S}$ to denote all the sentences previously derived, and assumptions which are in scope.

$\vee$–ELIMINATION

$$\frac{\mathcal{S},\mathcal{A}\vdash\mathcal{C},\quad \mathcal{S},\mathcal{B}\vdash\mathcal{C},\quad \mathcal{A}\vee\mathcal{B}}{\mathcal{C}}$$ where $\mathcal{A}$ and $\mathcal{B}$ are assumptions.

$\Rightarrow$–INTRODUCTION

$$\frac{\mathcal{S},\mathcal{A}\vdash\mathcal{B}}{\mathcal{A}\Rightarrow\mathcal{B}}$$ where $\mathcal{A}$ is an assumption.

$\neg$–INTRODUCTION

$$\frac{\mathcal{S},\mathcal{A}\vdash\mathcal{B},\quad \mathcal{S},\mathcal{A}\vdash\neg\mathcal{B}}{\neg\mathcal{A}}$$ where $\mathcal{A}$ is an assumption.

Derivation 2.4 *Show that*

$$P \Rightarrow Q\,,\; Q \Rightarrow R \vdash P \Rightarrow R$$

Derivation

1	$P \Rightarrow Q$	premise	
2	$Q \Rightarrow R$	premise	
3	P	assumption	|
4	Q	1,3 $\Rightarrow$–elimination	|
5	R	2,4 $\Rightarrow$–elimination	|
6	$P \Rightarrow R$	3,4,5 $\Rightarrow$–introduction	

QED

In this example it is important to realise that we have not derived the sentences Q and R on lines 4 and 5, since they both come within the scope of the assumption of P on line 3. Line 6, however, does contain a derived sentence, since $P \Rightarrow R$ comes from the inference rule which motivated the assumption being made in the first place, and hence it discharges the assumption at the preceding line.

Theorem 2.1 *Show that*

$$\vdash (P \wedge Q) \vee (\neg P \wedge R) \Rightarrow (Q \vee R)$$

Proof

1	$(P \wedge Q) \vee (\neg P \wedge R)$	assumption
2	$P \wedge Q$	assumption
3	Q	2 $\wedge$–elimination
4	$Q \vee R$	3 $\vee$–introduction
5	$\neg P \wedge R$	assumption
6	R	5 $\wedge$–elimination
7	$Q \vee R$	6 $\vee$–introduction
8	$Q \vee R$	2–7 $\vee$–elimination
9	$(P \wedge Q) \vee (\neg P \wedge R) \Rightarrow (Q \vee R)$	1–9 $\Rightarrow$–introduction

QED

In case you are mystified as to the strategy that was adopted in this proof, here is a brief explanation. The theorem to be proved is of the form $\mathcal{A} \Rightarrow \mathcal{B}$, so the strategy is to assume $\mathcal{A}$ and try to deduce $\mathcal{B}$; this is line 1. We are then faced with a sentence of the form $\mathcal{A} \vee \mathcal{B}$. The strategy here is to assume each side in turn, lines 2 and 5, and seek to derive a common result from each assumption. This we manage to do, lines 4 and 7. Having derived this common result from both assumptions, we can write it down as derived, and discharge the assumptions, line 8. We can then introduce the $\Rightarrow$, and discharge the outstanding assumption, line 9. You should note carefully that line 9 is the only theorem proved here, as every other line is within the scope of at least one assumption.

Derivation 2.5 *Show that*

$$P \Rightarrow Q \vdash \neg(P \land \neg Q)$$

Derivation

1	$P \Rightarrow Q$	premise
2	$P \land \neg Q$	assumption
3	P	2 $\land$–elimination
4	Q	1 $\Rightarrow$–elimination
5	$\neg Q$	2 $\land$–elimination
6	$\neg(P \land \neg Q)$	2-5 $\neg$–introduction

QED

This way of proceeding is sometimes called *proof by contradiction.* The strategy involves assuming the negation of that which is to be proved, and then showing that a contradiction arises, in this case lines 4 and 5. The $\neg$–introduction rule allows us to discharge the assumption, by introducing its negation. One important property that is related to this style of proof is that from an inconsistent set of premises (ie. one that contains both $\mathcal{A}$ and $\neg\mathcal{A}$ in some form) we can prove anything.

Derivation 2.6 *Show that*

$$P, \neg P \vdash Q$$

Derivation

1	$\neg Q$	assumption
2	P	premise
3	$\neg P$	premise
4	$\neg\neg Q$	1–3 $\neg$–introduction
5	Q	4 $\neg$–elimination

QED

This derivation captures arguments like "The moon is green and the moon is not green, so China is a suburb of London".

Exercise 2.9 *Show the following.*

1. $\vdash (P \vee P) \Leftrightarrow P$
2. $\vdash (P \wedge Q) \Rightarrow Q$
3. $\neg P \Rightarrow \neg Q\,,\, P \Rightarrow \neg R\,,\, R \vdash \neg(R \wedge Q)$
4. $Q \Rightarrow \neg P\,,\, P \wedge Q \vdash R$

Our next example shows how we can prove formally a property which is intuitively obvious, since it is just a formal statement of the law of the excluded middle. Unfortunately this "obvious" property has a proof which is certainly not obvious, but is a good test of understanding. Line 4 of this proof is not strictly necessary, but it exposes the contradiction more clearly.

Theorem 2.2 *Show that*

$$\vdash P \vee \neg P$$

Proof

1	$\neg(P \vee \neg P)$	assumption
2	P	assumption
3	$P \vee \neg P$	2 $\vee$–introduction
4	$\neg(P \vee \neg P)$	1 copied
5	$\neg P$	2–4 $\neg$–introduction
6	$P \vee \neg P$	5 $\vee$–introduction
7	$\neg\neg(P \vee \neg P)$	1–6 $\neg$–introduction
8	$P \vee \neg P$	7 $\neg$–elimination

QED

It should come as no great surprise that many other proofs and derivations make use of this property, but we would rather not have to remember how to prove it every time. For this reason we will assume a rule for theorem introduction, TI, which allows us to write down as a line in a derivation or proof any previously proved theorem. Our justification for this should be obvious, for we can always replace the inserted theorem by the text of its proof.

Derivation 2.7 *Show that*

$$P \Rightarrow Q \vdash \neg P \vee Q$$

Derivation

1	$P \Rightarrow Q$	premise
2	$P \vee \neg P$	TI theorem 2.2
3	P	assumption
4	Q	1,3 $\Rightarrow$–elimination
5	$\neg P \vee Q$	4 $\vee$–introduction
6	$\neg P$	assumption
7	$\neg P \vee Q$	6 $\vee$–introduction
8	$\neg P \vee Q$	2–7 $\vee$–elimination

QED

Proofs and derivations take place without consideration of the meanings of the proposition symbols being used. Consequently we will feel justified in introducing theorems with different representations for the propositions than those used in the presentation of the theorem, provided any substitution is performed consistently. This may include introducing sentences denoting compound propositions to replace simple ones in the theorem. We might introduce as an instance of Theorem 2.2, for example,

$$\vdash (P \Rightarrow (Q \vee \neg R)) \vee \neg(P \Rightarrow (Q \vee \neg R))$$

as this is still of the form $\mathcal{A} \vee \neg \mathcal{A}$.

We can also use previously performed derivations in this way. If we have performed a derivation of the form $\mathcal{S} \vdash \mathcal{W}$, and we are attempting another derivation in which we we have already obtained all the sentences of $\mathcal{S}$, then we can write down $\mathcal{W}$ without reproducing the entire derivation. Results introduced into proofs and derivations are called *lemmas*. These may be proved elsewhere, or taken for granted without a formal proof. Remember that your proofs and derivations are only as reliable as the weakest link, and leaving lemmas unproved is a notorious way of introducing errors.

Derivation 2.8 *Given, as a lemma, the derivation*

$$P \Rightarrow Q \vdash \neg P \vee Q$$

show that

$$(R \vee S) \Rightarrow (T \Leftrightarrow V) \vdash (\neg(R \vee S) \vee (T \Leftrightarrow V)) \vee W$$

Derivation

1	$(R \vee S) \Rightarrow (T \Leftrightarrow V)$	premise
2	$\neg(R \vee S) \vee (T \Leftrightarrow V)$	given derivation $R \vee S$ replaces P $T \Leftrightarrow V$ replaces Q
3	$(\neg(R \vee S) \vee (T \Leftrightarrow V)) \vee W$	2 $\vee$–introduction

QED

2.7 Consistency and Completeness

Propositional calculus has two properties which make it a very useful formal system: it is consistent and complete. These properties (or metatheorems about the calculus) can be stated as follows:

> If $\mathcal{P}$ denotes a set of sentences, then a formal system is complete, with respect to an interpretation, if whenever $\mathcal{P} \models \mathcal{W}$ then $\mathcal{P} \vdash \mathcal{W}$. That is to say if we can reason informally about the interpretations of $\mathcal{P}$and $\mathcal{W}$, then we can mirror the reasoning formally, that is purely syntactically, within the formal system.

and

> If $\mathcal{P}$ denotes a set of sentences, then a formal system is consistent with respect to an interpretation if whenever $\mathcal{P} \vdash \mathcal{W}$ then $\mathcal{P} \models \mathcal{W}$. That is to say if we can derive something formally then we can reason about the meanings of the sentences concerned and arrive at the same conclusion.

The consistency property gives us an explanation of what a derivation in propositional calculus means. If we can perform the derivation $\mathcal{P} \vdash \mathcal{W}$ then we know that $\mathcal{P} \models \mathcal{W}$, and so if all the sentences in $\mathcal{P}$ are given interpretations of true then $\mathcal{W}$ is also true. Our formal system gives us a syntactic way to model the deduction of true things from other true things. Derivation 2.6 showed that from an inconsistent set of premises we can derive anything. This can now be explained by observing that it models a deduction from a set of inconsistent facts (ie. they cannot all be true), so we should not be surprised if it produces strange conclusions.

This leads to the important heuristic result that we usually only write down things in derivations which we expect to be true in some interpretation, except for assumptions, where we might intentionally introduce something we expect to be false in order to derive a contradiction, as in Example 2.5.

2.8 Reducing Formality

The completeness and consistency results allow us to switch between the purely syntactic world of derivations and proofs, and the semantic ideas of truth and falsity. Equivalences, for example, can be used by observing the following

> If $\mathcal{A} \equiv \mathcal{B}$ then $\mathcal{A} \models \mathcal{B}$ and $\mathcal{B} \models \mathcal{A}$, and so, using our completeness property, $\mathcal{A} \vdash \mathcal{B}$ and $\mathcal{B} \vdash \mathcal{A}$.

Thus for every equivalence there are two corresponding derivations we know can be done, even if we don't know how to do them. This means that we can take short cuts in derivations and proofs by using equivalences, in the *well-founded* knowledge that there is a formal way of achieving the same result.

Theorem 2.3 *Given, as lemmas, the equivalences*

$$\mathcal{A} \Rightarrow \mathcal{B} \equiv \neg\mathcal{A} \vee \mathcal{B} \qquad \textit{and} \qquad \neg(\mathcal{A} \vee \mathcal{B}) \equiv \neg\mathcal{A} \wedge \neg\mathcal{B}$$

show that

$$\vdash (P \Rightarrow Q) \vee (Q \Rightarrow P)$$

Proof

1	$(P \Rightarrow Q) \vee \neg(P \Rightarrow Q)$	TI. theorem 2.2
2	$P \Rightarrow Q$	assumption
3	$(P \Rightarrow Q) \vee (Q \Rightarrow P)$	2 $\vee$–introduction
4	$\neg(P \Rightarrow Q)$	assumption
5	$\neg(\neg P \vee Q))$	4 by given lemma
6	$\neg\neg P \wedge \neg Q$	5 by given lemma
7	$\neg Q$	6 $\wedge$–elimination
8	$\neg Q \vee P$	7 $\vee$–introduction
9	$Q \Rightarrow P$	8 by given lemma
10	$(P \Rightarrow Q) \vee (Q \Rightarrow P)$	$\vee$–introduction
11	$(P \Rightarrow Q) \vee (Q \Rightarrow P)$	1–10 $\vee$–elimination

QED

We can avoid drawing up large truth tables to check for tautologies, for example, by noting that

If $\vdash \mathcal{A}$ then $\models \mathcal{A}$

and so we can choose to prove a theorem rather than investigate all possible interpretations of a complex sentence.

Exercise 2.10 *Show that* $\models P \vee \neg P \vee Q \vee R \vee S \vee T$ *by proving*

$$\vdash P \vee \neg P \vee Q \vee R \vee S \vee T$$

The inference rules can now be viewed as *common sense* techniques, rather than mystical symbolic rituals. The rule for removing $\vee$ signs, for example, can be explained as follows.

To investigate the consequences of $\mathcal{A} \vee \mathcal{B}$ being true, we can consider each side of the disjunction in turn. If we can find a common result which follows from both sides, then that result must follow from the disjunction, since at least one side must be true.

Exercise 2.11 *Investigate the other inference rules and see if you can find similar ways of explaining them.*

A useful relationship between proofs and derivations can be expressed in terms of $\Rightarrow$.

If $\mathcal{A} \vdash \mathcal{B}$ then $\vdash \mathcal{A} \Rightarrow \mathcal{B}$ and if $\vdash \mathcal{A} \Rightarrow \mathcal{B}$ then $\mathcal{A} \vdash \mathcal{B}$.

This can be seen by observing that a derivation showing $\mathcal{A} \vdash \mathcal{B}$ takes $\mathcal{A}$ as a premise, and a proof of $\vdash \mathcal{A} \Rightarrow \mathcal{B}$ starts with an assumption of $\mathcal{A}$. If the commentary were omitted then these would be indistinguishable, except for the last line of the proof which would introduce $\mathcal{A} \Rightarrow \mathcal{B}$. We will allow ourselves the convenience of introducing the theorem $\vdash \mathcal{A} \Rightarrow \mathcal{B}$ whenever we have performed the derivation $\mathcal{A} \vdash \mathcal{B}$.

Exercise 2.12 *Generalise this result to find a theorem corresponding to a derivation of the form* $\mathcal{A}, \mathcal{B}, \mathcal{C}, \ldots \vdash \mathcal{W}$.

Another useful way of introducing a sentence of the form $\mathcal{A} \Rightarrow \mathcal{B}$, which we will have occasion to use later in the book, arises from the following derivation.

Derivation 2.9 (Vacuous $\Rightarrow$ Introduction) *Show that*

$$\neg P \vdash P \Rightarrow Q$$

Derivation

1	$\neg P$	premise
2	P	assumption
3	$\neg Q$	assumption
4	$\neg\neg Q$	1-3 $\neg$–introduction
5	Q	4 $\neg$–elimination
6	$P \Rightarrow Q$	2-5 $\Rightarrow$–introduction

QED

This result allows us to introduce into a derivation or proof a sentence of the form $\mathcal{A} \Rightarrow \mathcal{B}$ whenever we have already written down a sentence of the form $\neg\mathcal{A}$, and the choice of $\mathcal{B}$ is entirely up to us.

The abilities to reduce formality in a well-founded manner and re-use previous results where appropriate are both fundamental to the ways in which mathematics proceeds. If every single result needed to be re-proved, from first principles and with total formality, before use then mathematics would never have developed into a useful tool. The rest of this book is devoted to presenting some mathematics in a way that shows it is well founded, but without providing total formality. As we progress we will build up a set of formal systems to which we will be able to assign useful interpretations in the world of software engineering.

Exercise 2.13 *Use any of the previously introduced theorems, derivations or equivalences to show the following: [Remember that to show* $\mathcal{A} \dashv\vdash \mathcal{B}$ *you must show both* $\mathcal{A} \vdash \mathcal{B}$ *and* $\mathcal{B} \vdash \mathcal{A}$*.]*

1. $P \wedge Q \vdash P \Rightarrow Q$
2. $\neg P \wedge \neg Q \vdash P \Rightarrow Q$
3. $\neg(Q \vee P) \vdash (Q \vee P) \Rightarrow ((P \wedge S) \Leftrightarrow (R \vee \neg Q))$
4. $\neg P \wedge \neg Q \vdash P \Leftrightarrow Q$
5. $(P \Rightarrow Q) \Rightarrow Q \dashv\vdash P \vee Q$
6. $P \vdash (P \wedge Q) \vee (P \wedge \neg Q)$
7. $(P \vee Q) \Leftrightarrow P \dashv\vdash Q \Rightarrow P$
8. $Q \vdash (P \wedge Q) \Leftrightarrow P$

2.9 Summary

In this chapter we have introduced a well-known example of a formal system, which has been developed to enable formal reasoning about propositions, a propositional calculus. We have discussed the properties of propositions, their classification according to truth values, and the way in which they may be regarded as simple or compound. The connectives provided by our formal system have been described, and truth tables have been introduced as a way of assigning meanings to sentences of the calculus.

The semantic turnstile has been introduced, and the relationship between formal, syntactic, derivations and semantic reasoning has been outlined. A natural deduction system forming the deductive apparatus of our formal system has been used for proofs and derivations, and ways of easing the burden of total formality have been introduced[4].

[4]For a more detailed account of propositional calculus, and predicate calculus covered in the next chapter, see W.H. Newton-Smith, *Logic-An Introductory Course*, Routledge & Kegan Paul, 1985, or E.J. Lemmon, *Beginning Logic*, Van Nostrand Reinhold(UK), 1965.

COMMUTATIVE PROPERTIES

$$\mathcal{A} \wedge \mathcal{B} \equiv \mathcal{B} \wedge \mathcal{A}$$

$$\mathcal{A} \vee \mathcal{B} \equiv \mathcal{B} \vee \mathcal{A}$$

ASSOCIATIVE PROPERTIES

$$(\mathcal{A} \wedge \mathcal{B}) \wedge \mathcal{C} \equiv \mathcal{A} \wedge (\mathcal{B} \wedge \mathcal{C})$$

$$(\mathcal{A} \vee \mathcal{B}) \vee \mathcal{C} \equiv \mathcal{A} \vee (\mathcal{B} \vee \mathcal{C})$$

DISTRIBUTIVE PROPERTIES

$$\mathcal{A} \vee (\mathcal{B} \wedge \mathcal{C}) \equiv (\mathcal{A} \vee \mathcal{B}) \wedge (\mathcal{A} \vee \mathcal{C})$$

$$\mathcal{A} \wedge (\mathcal{B} \vee \mathcal{C}) \equiv (\mathcal{A} \wedge \mathcal{B}) \vee (\mathcal{A} \wedge \mathcal{C})$$

DE MORGAN'S RULES

$$\neg(\mathcal{A} \vee \mathcal{B}) \equiv \neg\mathcal{A} \wedge \neg\mathcal{B}$$

$$\neg(\mathcal{A} \wedge \mathcal{B}) \equiv \neg\mathcal{A} \vee \neg\mathcal{B}$$

DOUBLE NEGATION PROPERTY

$$\neg\neg\mathcal{A} \equiv \mathcal{A}$$

Figure 2.1: Common Equivalences of Propositional Logic.

Chapter 3

Predicate Calculus

In this chapter we are going to explore another formal system, called predicate calculus. This formalism allows us to reason about statements which are rather more complex than propositions, but are similar in nature. It will be built upon propositional calculus by adding more expressive power to the language and additional rules to the deductive apparatus. This formal system will then be enriched to provide the notion of equality.

3.1 Predicates

In the previous chapter we presented a formalism for capturing the truth values of propositions. Simple propositions were denoted by letters, and compound forms by sentences using the connectives. This formalism is very limited in the type of arguments that can be captured, as we can see by considering the following example

> Esmerelda is a duck, ducks like water so Esmerelda likes water.

This argument uses three simple propositions which propositional logic can capture as P, Q, and R. Once we have made this abstraction, however, it is obvious that we will not be able to capture the argument, because our deductive apparatus contains no rules to allow a derivation of R from P and Q (and quite rightly, or we could deduce any result we liked from any arbitrary pair of propositions).

In order to formalise arguments like the above, we need to find a formalism which allows us to retain sufficient of the structure of the statements while freeing us from the details of a particular means of expression. This is achieved by introducing the idea of a *predicate.*

In the above example we really want to capture Esmerelda's "duckness" in a way which avoids the problem of representations. We do not mind if this "duckness" is pronounced by

- Esmerelda is a duck
- Esmerelda is a member of the duck family
- she is a duck, that Esmerelda.

Our formalism will allow us to capture this by providing some components in the formal language which map on to the properties of objects. These are called *predicates.* We will choose to denote predicates by names which reflect the property they are intended to capture, rather than limit ourselves to names such as P, Q, or R. The property of being a duck, for example, can be captured by the predicate, *duck*(x), where the x is a *free variable*, or *place holder*, which can be filled in by the names of suitable objects to create propositions. We can write, for example, *duck*(*Fred*) which will be true if *Fred* is a duck, but false if *Fred* is a dog. If the result of such instantiations are to be propositions, they must conform to the laws of contradiction and the excluded middle. We need to be aware, therefore, of the problems that may arise by instantiating predicates with names which give rise to funny cases like *duck*(*love*), since few of us would know how to discuss the duckness of love, and most of us, philosophers excepted, would rather not bother!

The predicate *duck*(x) is called a *unary*, or *one-place*, predicate since it has just one place for a name to be put. The choice of our free variable in a predicate is arbitrary as it is just a place holder. The predicate $P(x)$ could equally well be written $P(y)$.

If we allow *n-place predicates* then we can capture the idea of some relationship between n objects. The predicate *father of*(x, y), for example, could be used to capture the idea that x is the father of y, and *relay team*(w, x, y, z) might denote the fact that four runners form a relay team. The predicate *numerically bigger than*(x, y) might be used to capture the idea that x is a larger number than y. A more conventional way might be to write $x > y$; this is still a predicate, but it is in *infix form*.

When we come to find interpretations we must remember that any names we have chosen to use in the formal system must be assigned meanings, even if the names themselves suggest what those meanings might be.

3.2 The Syntax of Predicate Logic

Here is the syntax of a formal language which we can use to capture the truth value of statements involving predicates and propositions. It is expressed in the standard syntactic metalanguage, but not completely formally, in Figure 3.1. We are not going to be precise in specifying symbols to denote predicates and objects, as with P, Q and R in propositional logic, rather we will allow ourselves to choose suitable names to stand for objects and predicates in the light of intended interpretations. We have indicated this freedom with comments in the grammar, denoted by "(*..........*)" in the metalanguage.

Exercise 3.1 *Which of the following are well formed sentences of predicate logic?*

1. $\forall x \bullet P(x)$
2. $\exists y \bullet Q$
3. $P \Rightarrow Q$
4. $\forall x \bullet \exists P \wedge Q$
5. $\forall x \bullet \exists yP(x, y)$
6. $\forall \exists xyP(x, x)$

We can see from this grammar that any sentence of propositional logic is also a sentence in predicate logic. Propositions may also be denoted by $P(t)$, where P is an n-ary predicate and t is a list of n proper or arbitrary names. These propositions are still truth valued, but they also capture additional information.

```
sentence             = simple proposition|predicate
                       |"¬", sentence
                       |"(", sentence, " ∧ ", sentence")"
                       |"(", sentence, " ∨ ", sentence")"
                       |"(", sentence, " ⇒ ", sentence")"
                       |"(", sentence, " ⇔ ", sentence")"
                       |" ∀ ", variablename, " • ", sentence
                       |" ∃ ", variablename, " • ", sentence;
simple proposition   = "P"|"Q"|"R" ...;
predicate            = predicate name, "(", termlist, ")";
predicate name       = ....(*These will be left free*);
term list            = term|term, ",", term list;
term                 = proper name|arbitrary name|variable name;
proper name          = ....(*These will be left free*);
arbitrary name       = "a"|"b"|"c" ...;
variable name        = "x"|"y"|"z" ...;
```

Figure 3.1: The Syntax of Predicate Logic

A predicate itself is *not* truth valued, it expresses a property or relation using variables. Predicates can give rise to propositions, however, in two ways. First, as we have already seen, their free variables may be instantiated with the names of objects, and second something can be said about the instantiation process itself using the technique of *quantification*. Quantification introduces the two additional symbols of our formal system, $\forall$ and $\exists$.

For a unary predicate, $P(x)$, with free variable x, both

$$\exists x \bullet P(x) \qquad \text{and} \qquad \forall x \bullet P(x)$$

are propositions. The variable x is now said to be *bound* by the quantification, and can no longer be instantiated. For a predicate with more than one free variable, $Q(x, y)$, quantifying over one of the free variables gives rise to another predicate, as one of the variables remains free. Thus from $Q(x, y)$ we can obtain for example

$$\forall x \bullet Q(x, y) \qquad \text{or} \qquad \exists y \bullet Q(x, y)$$

These predicates can now be quantified over the remaining variables, giving, for example,

$$\forall y \bullet \forall x \bullet Q(x, y) \qquad \text{or} \qquad \forall x \bullet \exists y \bullet Q(x, y)$$

Our choice of free variable name in a predicate is irrelevant, of course, and so

$$\forall x \bullet P(x) \qquad \text{and} \qquad \forall y \bullet P(y)$$

denote the same proposition. This might lead you to believe that bound variables in propositions can be renamed at will, but care must be taken not to introduce confusion by giving two distinct free variables the same name. We will not allow, for example,

$$\exists x \bullet \forall x \bullet P(x, x)$$

to arise as it is unclear which quantification binds which variable. Similarly we will not allow quantification such as

$$\forall x \bullet P(y)$$

to occur[1].

3.3 Giving Predicate Logic a Semantics

The conventional interpretation of predicate logic is rather more complex than for propositional logic, as the language itself is more powerful. We interpret any simple propositional symbols and the connectives as in propositional logic and assign any proper names used to

[1]Strictly the language admits both of these forms, but they are usually treated as meaningless when semantics are given. We will attempt to avoid introducing forms which we know in advance are meaningless.

objects in the real world[2]. Bound variables, in quantified expressions, are *not* assigned any meaning.

We allow our predicate symbols to denote properties of objects or relations between them, taking care that the predicate has the correct number of free variables for its intended meaning. Using $likes(x, y)$ to denote the property "x is a duck" is not allowed, since it leaves us wondering what rôle the free variable y plays.

There are two new symbols which need to be explored further: $\forall$ and $\exists$. These are used to capture the idea of *universal* and *existential* quantification respectively. Universal quantification allows us to write propositions of the form "every object has this property" or "all objects are related in this way". They are propositions because they have an underlying truth value and conform to the rules of excluded middle and contradiction, but they are expressed in terms of predicates. A sentence of the form $\forall x \bullet P(x)$ is interpreted as the proposition "we can instantiate x by the name of any object and the resulting proposition will be true". This assertion may itself be true or false, of course.

Example 3.1 $\forall x \bullet duck(x)$ *captures the assertion that "every object is a duck".*

There is a complication to this, however, namely that some objects may not really be capable of being used meaningfully in some predicates.

Consider the predicate $greater\ than(x, y)$ which we are going to use to mean the property that x is numerically greater than y. Under any sensible interpretation $greater\ than(7, 4)$ will be true, but what will $greater\ than(cup\ of\ coffee, Fred)$ mean? The concept, of being numerically greater than, clearly does not apply, so it is not sensible to interpret this as true or false, and thus the sentence is not a proposition since it has a middle value. If we universally quantify some expression there are almost certainly some objects within the collection of all the objects in the universe for which this problem will arise.

---

[2] If we chose terms like "Fred" then this may seem a trivial step, but remember that in theory we could allow "Fred" to denote the person Bill if we really wanted to be contrary.

This is a very deep problem in logic which we will overcome by introducing the idea of a *domain of interest* which contains precisely those objects for which the predicate will make sense, and then considering quantification only over all the values in this domain. In a later chapter we will see how, with a little more notation, we could extend our formal system to make this explicit and more powerful.

A sentence of the form $\exists x \bullet P(x)$ asserts that there is at least one value in the domain of interest for which the predicate $P(x)$ is true. This can be used to capture notions like

"there is at least one duck" as $\exists x \bullet duck(x)$

"Bill has a father" as $\exists x \bullet father\ of(x, Bill)$

Note that $\exists x \bullet father\ of(x, Bill)$ does not really capture "Bill has a father" perfectly, rather it captures "Bill has at least one father". An extended version of predicate logic might well give us a new symbol, typically "$\exists!$" or $\exists_1$, to denote "there exists exactly one".

We can now look at how the truth value of some compound forms of English expression might be captured in predicate logic. We will not attempt to discuss this in depth, but will illustrate the idea with some simple examples and leave the philosophical issues to further reading. For our first example let us explore how we might capture

ducks like the pond

where the pond is an identifiable object in our domain of interest. First we need to formalise our language so it has terms and predicates corresponding to the objects and properties being discussed. We will choose $duck(x)$ to denote the assertion that some object is a duck, *pond* to represent the pond and $likes(x, y)$ to denote x likes y.

Next we must think carefully about what we are trying to say. This sentence could be re-phrased as "whatever value we choose for x, if x is a duck then x likes the pond". The "if $x \ldots$ then $\ldots$" construct has been encountered before, and is reflected by the $\Rightarrow$ symbol in propositional logic.

This sort of reasoning leads us to a sentence of the form

$$\forall x \bullet (duck(x) \Rightarrow likes(x, pond))$$

Note that if there are no ducks in our domain of interest then $duck(x)$ will never be instantiated with an object which makes it true. In this case the definition of $\Rightarrow$ is such that the above expression will be true.

As another example let us try capturing

there is a duck that likes the pond.

One major difference between this example and the last one is that now we are trying to express the fact that there is at least one duck in our universe of discourse. It is now not appropriate to think in terms of "if there is a duck then ..." but of "there is a duck and ...". With the benefit of this insight we can now see that the sentence

$$\exists\, x \bullet (duck(x) \wedge likes(x, pond))$$

captures this idea; there is an object in our domain of interest such that this object is a duck and it likes the pond.

For a more complicated example, let us express

ducks like water

using predicate logic. Here we need to say that not only do ducks like the pond, but they like any object which has the property of being water. It is worth observing that when we explicitly named "the pond" we were assuming its existence, but now we are not sure that there is any object which is watery; neither are we sure that there are any ducks in our universe.

If we use the predicate $water(y)$ to denote that y has the property of being water, then we want to say that "if x is a duck and y is a watery object then x likes y", and this is true whatever objects we choose for x and y. Thus we can write

$$\forall\, x \bullet \forall\, y \bullet (\text{if } x \text{ is a duck and } y \text{ is a watery object then } x \text{ likes } y)$$

or

$$\forall\, x \bullet \forall\, y \bullet ((duck(x) \wedge water(y)) \Rightarrow likes(x, y))$$

This will be true if there are no ducks, or if there is no water.

Exercise 3.2 *Express the above example so that it also asserts that there is some water somewhere in the world.*

As a final example let us capture

> there is a duck who likes every watery object.

This is an ambiguous statement in English, because it could mean that there is one specific, water-loving, duck who likes every watery object or it could mean that for each watery object there is a duck somewhere that likes it. The first case can be rephrased as

> There is at least one object which is a duck and (the same object) likes every watery object.

or

$$\exists\, x \bullet (x \text{ is a duck and } x \text{ likes every watery object})$$

"x likes every watery object" can be expressed by

$$\forall\, y \bullet \mathit{watery}(y) \Rightarrow \mathit{likes}(x, y)$$

so we can capture the whole expression as

$$\exists\, x \bullet (\mathit{duck}(x) \wedge (\forall\, y \bullet \mathit{watery}(y) \Rightarrow \mathit{likes}(x, y)))$$

Exercise 3.3 *Express the second possible meaning of the statement in predicate logic.*

The semantic concept of validity introduced for propositional logic can now be extended to interpretations of predicate logic. Showing validity of expressions in predicate logic is much harder, however, because we have to consider all possible assignments of meanings to terms and predicates. Moreover, if we want to discuss the validity of something which is universally quantified then we need to show that it is true for all objects in the domain of discourse, and this domain may be infinite. Truth tables will not work for predicate logic, and there is no similar device which can be generally used to show validity. Fortunately we can find a deductive apparatus for predicate calculus which is both consistent and complete, with respect to the class of interpretations we have outlined above.

There are some useful equivalences in predicate logic which we have summarised below; as before these are not going to form part of our formal system, but the completeness property allows us to utilise them in a well-founded way.

$\forall x \bullet P(x) \equiv \neg \exists x \bullet \neg P(x)$ [all true—none false]

$\forall x \bullet \neg P(x) \equiv \neg \exists x \bullet P(x)$ [all false—none true]

$\neg \forall x \bullet P(x) \equiv \exists x \bullet \neg P(x)$ [not all true—at least one false]

$\neg \forall x \bullet \neg P(x) \equiv \exists x \bullet P(x)$ [not all false—at least one true]

3.4 Predicate Calculus

The deductive apparatus is built upon that for propositional calculus by adding four new rules for the introduction and elimination of the two quantifiers. Rather than treat these rules as purely syntactic, which of course they are, we will look at what they mean in terms of the use to which the formal system will be put. We can do this because we already know that we have the consistency and completeness properties.

Consider the sentence $\forall x \bullet P(x)$, and let us suppose we know that the term *Fred* interprets to some object in our universe of discourse. Since we interpret $\forall x \bullet P(x)$ as "$P(x)$ is true for all objects" we are implicitly asserting $P(Fred)$. This gives us a rule for eliminating the universal quantifier, but it is rather a weak one; we have made a particular choice of *Fred* as the object. A stronger rule would allow us say that, if the names of the objects in the domain of discourse are $n_1, n_2, \ldots, n_k$, then from $\forall x \bullet P(x)$ we can infer

$$P(n_1) \wedge P(n_2) \wedge \cdots \wedge P(n_k)$$

but this will be a very large sentence, especially if the domain is infinite!

We overcome this problem by introducing the idea of an *arbitrary term* into derivations. This is a term which can stand for any one of the objects in the universe of discourse, chosen without any imposed constraints. In our derivations we will use the letters $a, b, \ldots$ to denote arbitrary terms. This gives us a rule for eliminating the universal quantifier.

∀–ELIMINATION

$$\frac{\forall x \bullet P(x)}{P(a)} \qquad \text{where } a \text{ is arbitrary.}$$

This suggests a way of introducing a universal quantifier, because if we know that some property is being asserted for objects chosen arbitrarily then it must be being asserted for every object in the domain of interest. There must be no constraints on the arbitrary objects, of course, and hence there must not be any assumptions currently being made about the arbitrary object chosen.

∀–INTRODUCTION

$$\frac{P(a)}{\forall x \bullet P(x)} \qquad \text{where } a \text{ is arbitrary.}$$

To eliminate a universal quantifier we do not have to introduce an arbitrary term, we could use a specific one, such as *Fred* above. To introduce a universal quantifier, however, we must have a predicate involving an arbitrary term, so it is important to remember whether a term is arbitrary or specific.

The problem of eliminating existential quantifiers is similar to that for universal elimination, except that from $\exists x \bullet P(x)$ we can infer

$$P(n_1) \vee P(n_2) \vee \cdots \vee P(n_k)$$

but it is not obvious how we should handle this large disjunction of terms. Observe, however, that if we have a sentence of the form $\exists x \bullet P(x)$, and we could show that $(\exists x \bullet P(x)) \Rightarrow Q$, then we could derive the sentence Q, which would have eliminated the existential quantification. Now showing that $(\exists x \bullet P(x)) \Rightarrow Q$ is in general quite hard, but we can make use of the following equivalence

$$\forall x \bullet (P(x) \Rightarrow Q) \equiv (\exists x \bullet P(x)) \Rightarrow Q$$

and derive the universally quantified form instead. This provides the basis of our rule for existential elimination.

∃–ELIMINATION

$$\frac{\exists x \bullet P(x) \quad , \quad \forall x \bullet (P(x) \Rightarrow Q)}{Q}$$

Exercise 3.4 *Show the above equivalence holds by considering a finite domain of interest with just two members, and performing a case analysis.*

Now, if we have a sentence of the form $P(t)$ then this asserts $P(x)$ to be true for some term t. This term may be the name of a real object or an arbitrary term introduced by a previous universal quantification elimination; in both cases it asserts that there is at least one object for which P holds true. Thus we have the rule

∃–INTRODUCTION

$$\frac{P(t)}{\exists x \bullet P(x)} \qquad \text{where } t \text{ is any term}$$

We now have the deductive apparatus for predicate calculus, and can attempt some derivations and proofs. We will follow our derivations with a discussion of the interpretations we have in mind, thus explaining the syntactic manipulations by giving an example of corresponding semantic reasoning.

Derivation 3.1 *Show that*

$$P(m), \forall x \bullet (P(x) \Rightarrow Q(x)) \vdash Q(m)$$

Derivation

1	$\forall x \bullet (P(x) \Rightarrow Q(x))$	premise
2	$P(m) \Rightarrow Q(m)$	1 ∀–elimination
3	$P(m)$	premise
4	$Q(m)$	2,3 ⇒–elimination

QED

If we give the following interpretation to the terms and predicates

$P(x)$ denotes x is a duck

$Q(x)$ denotes x likes the pond

m denotes Esmerelda

then $\forall x \bullet (P(x) \Rightarrow Q(x))$ denotes "all ducks like the pond", $P(m)$ denotes "Esmerelda is a duck" and the derivation allows us to arrive at $Q(m)$ which denotes "Esmerelda likes the pond". Thus we have managed to handle the type of argument that propositional calculus was unable to capture.

Derivation 3.2 *Show that*

$$\forall x \bullet P(x) \vdash \forall x \bullet (Q(x) \Rightarrow P(x))$$

Derivation

1	$\forall x \bullet P(x)$	premise
2	$Q(a)$	assumption, arbitary a
3	$P(a)$	1 $\forall$–elimination
4	$Q(a) \Rightarrow P(a)$	2,3 $\Rightarrow$–introduction
5	$\forall x \bullet (Q(x) \Rightarrow P(x))$	4 $\forall$–introduction

QED

This example can be explained without the need to give a particular interpretation, since it shows that if the right-hand side of the construct $\mathcal{A} \Rightarrow \mathcal{B}$ is asserted to be true then the whole construct is asserted to be true regardless of the left-hand side.

Exercise 3.5 *Convince yourself of this fact by reviewing the truth table for* $\mathcal{A} \Rightarrow \mathcal{B}$.

Derivation 3.3 *Show that*

$$\forall x \bullet \forall y \bullet P(x,y) \Rightarrow \neg P(y,x) \vdash \forall x \bullet \neg P(x,x)$$

Derivation

1	$\forall x \bullet \forall y \bullet P(x,y) \Rightarrow \neg P(y,x)$	premise
2	$\forall y \bullet P(a,y) \Rightarrow \neg P(y,a)$	1 $\forall$–elimination
3	$P(a,a) \Rightarrow \neg P(a,a)$	2 $\forall$–elimination
4	$P(a,a)$	assumption
5	$\neg P(a,a)$	3,4 $\Rightarrow$–elimination
6	$\neg P(a,a)$	4,5 $\neg$–introduction
7	$\forall x \bullet \neg P(x,x)$	6 $\forall$–introduction

QED

In this example, our choice of arbitrary name on line 3 is important. Although the rule allows a totally free choice of arbitrary or particular name, we need to guide this choice using the context of the problem. In this case, using different arbitrary names for the two eliminations would not have allowed us to achieve the desired result.

Derivation 3.3 is the kind of general result upon which mathematics is based. It says something about the properties of relations between objects. For example, if we let $P(x, y)$ denote the "less than" relation of conventional arithmetic, $x < y$, then

$$\forall x \bullet \forall y \bullet P(x,y) \Rightarrow \neg P(y,x) \quad \text{means} \quad (x < y) \Rightarrow \neg(y < x)$$

This being the case, the derivation allows us to conclude that there is no value of x for which $x < x$.

This property is not true of just the one interpretation we have given, however, it is true for any interpretation for which the premises hold true. It forms part of the *theory of relations* which will be discussed in Chapter 6. In the terminology of theory of relations, we would say that we have proved that *all asymmetric relations are irreflexive.*

Derivation 3.4 *Show that*

$$\forall x \bullet (P(x) \Rightarrow Q(x)),\ \exists y \bullet P(y) \vdash \exists z \bullet Q(z)$$

Derivation

1	$\forall x \bullet P(x) \Rightarrow Q(x)$	premise
2	$\exists y \bullet P(y)$	premise
3	$P(a) \Rightarrow Q(a)$	1 $\forall$–elimination
4	$P(a)$	assumption
5	$Q(a)$	3,4 $\Rightarrow$–elimination
6	$\exists z \bullet Q(z)$	5 $\exists$–introduction
7	$P(a) \Rightarrow \exists z \bullet Q(z)$	4–6 $\Rightarrow$–introduction
8	$\forall x \bullet (P(x) \Rightarrow \exists z \bullet Q(z))$	7 $\forall$–introduction
9	$\exists z \bullet Q(z)$	2,8 $\exists$–elimination

QED

In this derivation we are applying the rule for existential elimination, so we need to derive the universal statement on line 8 first. Note that in this example the statement Q that we are trying to produce from the existential elimination is itself an existentially quantified form, but it is still a proposition of course.

Exercise 3.6 *Show the following*

1. $\forall x \bullet P(x) \dashv\vdash \forall y \bullet P(y)$
2. $\forall x \bullet P(x) \vdash \exists y \bullet P(y)$
3. $\forall x \bullet P(x) \vee Q(x) \dashv\vdash \forall x \bullet P(x) \wedge \forall y \bullet Q(y)$
4. $\neg \forall x \bullet \neg P(x) \dashv\vdash \exists x \bullet P(x)$
5. $\forall x \bullet P \wedge Q(x) \dashv\vdash P \wedge \forall x \bullet Q(x)$
6. $\forall x \bullet \forall y \bullet P(x, y) \dashv\vdash \forall y \bullet \forall x \bullet P(x, y)$
7. $\forall x \bullet P(x) \Rightarrow Q(x), \neg Q(a) \vdash \neg P(a)$
8. $\vdash \forall x \bullet P(x) \vee \exists y \bullet \neg P(y)$
9. $P(a) \Rightarrow \forall x \bullet \neg P(x) \vdash \neg P(a)$

3.5 Equality

In this final section we will add one more component to predicate logic, to provide us with a formal system powerful enough to be genuinely useful in software engineering. Frequently we wish to give the same object several names in different contexts, while still acknowledging that there is only one object.

Example 3.2 *Here are four ways of naming the same thing*

- *the number six*
- $3 + 3$
- *the smallest whole number greater than five*
- *the positive square root of* 36

This presents a problem to predicate logic as we have introduced it so far, as we would like to be able to refer to objects by whatever name seems appropriate at the time, but then to reason about these sentences in a way which recognises that several terms are all synonyms. The solution is simple: we must provide a rule which allows us to substitute equivalent names into a sentence without changing its truth value. To achieve this we first extend the formal language by providing a construction to denote the fact that two names refer to the same object. We could use, for example, a special predicate name $equals(x, y)$ to have the interpretation that x and y both denote the same object. There is, however, a more conventional notation, that of the infix predicate " $=$ ". You will recall that we carefully avoided using this symbol to denote equivalence, since we wanted equivalence to be a statement *about* some sentences in our logic, whereas equality is a concept we are going to capture *within* our formal system. The statement $\mathcal{A} \equiv \mathcal{B}$ says that the sentence $\mathcal{A}$ really does have the same truth value as the sentence $\mathcal{B}$. The assertion $p = q$, however, is a sentence in our formal system which may be either true or false.

To handle proofs and derivations which involve this special predicate we will introduce some additions to our deductive apparatus. First a rather simple axiom, which should surprise no-one.

AXIOM OF REFLEXIVITY

$$\forall x \bullet x = x$$

This is really just a statement of the fact that we cannot use the same name to denote two different objects. We will allow ourselves the luxury of using this rule for particular terms without always having to introduce the quantified form and then removing the quantifier, thus we will write down $n = n$ directly if we need it. This is sometimes referred to as an *axiom schema* since we can view the quantified form as a schema, or template, from which we can generate an infinite number of other axioms by instantiation with particular names.

The second thing we will add is a derivation rule. If we already have a sentence of the form $m = n$ then we can substitute m for occurrences of n in any preceding sentences in the derivation. To express this formally we need to add some power to our meta-language: we will use $S(n)$ to denote any sentence in predicate logic which involves n as a term *but not a bound variable* and $S[m/n]$ to denote the same sentence as S but with some of the occurrences of n replaced by m.

SUBSTITUTION RULE

$$\frac{m = n, S(n)}{S[m/n]} \quad \text{and} \quad \frac{m = n, S(m)}{S[n/m]}$$

We can now perform some derivations.

Derivation 3.5 *Show that*

$$s = t \vdash t = s$$

Derivation

1	$s = t$	premise
2	$s = s$	reflexivity
3	$t = s$	1,2 substitution (replacing first s in 2 by t)

QED

In spirit with our desire to re-use useful results, we will use this very simple, but important, result in future derivations by citing "symmetry".

Derivation 3.6 *Show that*

$$s = t \vdash (s = u) \Rightarrow (t = u)$$

Derivation

1	$s = t$	premise
2	$s = u$	assumption
3	$t = u$	1,2 substitution
4	$(s = u) \Rightarrow (t = u)$	2,3 $\Rightarrow$–introduction

QED

Exercise 3.7 *Show the following*

1. $s = t\,,\, t = u \vdash s = u$
2. $(a = b) \wedge (b = c)\,,\, P(a) \vdash P(b) \vee P(c)$

3. $(a = b) \vee (a = c)\,,\ P(b) \Rightarrow \neg P(c)\ ,\ P(a) \vdash \neg(b = c)$

We can use equality to capture the idea that a predicate $P(x)$ holds for at most one object.

$$\forall\, x \bullet \forall\, y \bullet ((P(x) \wedge P(y)) \Rightarrow (x = y))$$

We can also say that P(x) holds for precisely one object.

$$\exists\, x \bullet (P(x) \wedge \forall\, y \bullet (P(y) \Rightarrow (x = y)))$$

Exercise 3.8 *Find expressions in predicate logic with equality to express the following.*

1. *P(x) holds for exactly two objects.*
2. *P(x) holds for at most two objects.*

3.6 Summary

In this chapter we have introduced a formal system suitable for capturing a wide class of arguments, predicate calculus. We have shown how predicates can be used to capture properties of objects and relations between them, and how propositions can be obtained from predicates by instantiation or quantification. The semantics of predicate logic have been discussed and interpretations of sentences containing quantification have been given. A deductive apparatus has been constructed which permits proofs and derivations.

The concept of equality has been formally introduced, and the deductive apparatus has been extended to allow reasoning about equality.

Chapter 4

Theories

In this chapter we will introduce the idea of a *theory*, a word which most of us have met in phrases such as "the theory of relativity" in physics, or "the theory of monetarism" in economics. We will explain how formal systems can be used to present certain kinds of theories, and give examples of theory presentations built from the mathematical ideas covered so far in the book.

4.1 Theory Presentations

Intuitively we can think of a theory as the set of all the statements we want to make about some phenomenon. In practice, most theories contain too many facts for them to be explicitly listed, so they are often presented as a subset of the facts together with some laws from which the remainder can be derived.

If we want to introduce some formality into the way we present a theory then we may be able to find a suitable formal system, in which the fundamental facts can be expressed as axioms and the laws captured within a deductive apparatus. It should be remembered that not all theories can be formalised, and we are not claiming that theories *are* formal, merely that there are occasions when it is both possible and useful to formalise them. If we do decide to give a presentation of a theory as a formal system, then the theory itself can be thought of as the *closure* of all the theorems provable within the system.

The theorems in our theory can be viewed as being of two distinct types, those that are true only because of specific details about some phenomenon, such as $E = MC^2$ in the general theory of relativity, and those that are tautologies because of some facet of the way we are thinking. For example, if we choose to view the statement "hydrogen is an inert gas" as a proposition, then the statement "either hydrogen is an inert gas or it is not an inert gas" is a tautology, by the law of the excluded middle. The truth of this statement is independent of the substance of its constituent parts, and sentences of this form will be true in every theory which uses the concept of propositions. For this reason it is conventional to split a theory presentation into two parts, the axioms and inference rules for the specific theory, and a logical calculus to underpin it, such as predicate calculus with equality.

The choice of underlying logic is important, since any properties of the logic will automatically become part of the theory. There are many occasions in software engineering when predicate logic with equality is not the most suitable choice, such as when discussing distributed systems, when it might be desired to allow some condition to be both true and false, but in different places or at different times. Space does not permit us to develop all of the logics which are currently being proposed as useful for underpinning theory presentations in various realms of software engineering, so we will be content to explain the principle with a vehicle of predicate logic with equality, secure in the belief that once you have been shown how to lay foundations you can continue to extend the basis of your well-founded knowledge.

Having decided upon an underlying logic for our theory presentation, we need to restrict the formal language to include only those terms and predicate names which will have interpretations in the theory. To illustrate this we will build a theory about the Smith family.

> There are three named male Smiths: John, Bill and Harry.
> Harry is John's father and John is Bill's father.

First we need to establish the language we are going to use, so we will name people in the following way.

$$\begin{aligned} person = {} & \text{"}Harry\text{"} \mid \text{"}Bill\text{"} \mid \text{"}John\text{"} \\ & \mid \text{"}father(\text{"}, person, \text{"})\text{"} \\ & \mid \text{"}grandfather(\text{"}, person, \text{"})\text{"}; \end{aligned}$$

We add to the normal inference system for predicate calculus the axioms for our specific theory.

Axiom 1 $father(John) = Harry$

Axiom 2 $father(Bill) = John$

Axiom 3 $\forall x \bullet father(father(x)) = grandfather(x)$

We can now prove the following theorem.

Theorem 4.1 *Show that*

$$\vdash Harry = grandfather(Bill)$$

Proof

1	$father(John) = Harry$	axiom
2	$father(Bill) = John$	axiom
3	$father(father(Bill)) = Harry$	1,2 substitution
4	$\forall x \bullet father(father(x)) = grandfather(x)$	axiom
5	$father(father(Bill)) = grandfather(Bill)$	4 $\forall$–elimination
6	$Harry = grandfather(Bill)$	3,5 substitution

QED

This theory is rather more powerful than you might think, as we are not restricted to discussing Harry, Bill and John, but can also discuss any of their male ancestors, although they can only be referred to by their relationship to one of our named individuals.

Exercise 4.1 *Prove the theorem*

$$\vdash father(father(father(John))) = grandfather(grandfather(Bill))$$

Since theories presented in this way are nothing more than formal systems they can be given more than one interpretation.

Example 4.1 *Another interpretation for the above theory is*

- *let Bill denote 2*
- *let John denote 4*
- *let Harry denote 8*
- *let father(x) denote* $2 \times x$
- *let grandfather(x) denote* $4 \times x$

Exercise 4.2 *Show that the axioms of the theory about the Smiths apply to this interpretation, and then give interpretations for the two theorems introduced above.*

Exercise 4.3 *Convince yourself that this theory is incomplete, by trying to prove the intuitively obvious fact that no-one is his own father.*

An interpretation of a theory in which all of the axioms of the theory interpret to true is called a *model of the theory*, and this is usually the only sort of interpretation we are interested in. When constructing a theory it is usual to have at least one model in mind.

In order to overcome problematic incompleteness in a theory we can *enrich* it by adding more axioms to plug the perceived gaps, thereby creating a new, more powerful, theory. We must take great care, however, that in so doing we do not make the resulting theory inconsistent.

As an example of enrichment, here is a new axiom for the above theory, which states that fathers are older than their sons. We will capture "older than" using the symbol ">" in infix form. Thus, for example, $Harry > Bill$ denotes the fact that Harry is older than Bill.

Axiom 4 $\forall x \bullet father(x) > x$

If you attempt to use this axiom you will probably discover that you need another one too, namely

Axiom 5 $\forall x \bullet \forall y \bullet \forall z \bullet (x > y) \wedge (y > z) \Rightarrow (x > z)$

This axiom contains three variables, all universally quantified, and is rather daunting; to simplify the notation you may often see it written as

$$\forall x, y, z \bullet (x > y) \wedge (y > z) \Rightarrow (x > z)$$

or even just

$$(x > y) \wedge (y > z) \Rightarrow (x > z)$$

where the quantification is taken for granted. This is another example of how formality is frequently reduced, but in a well understood manner.

We can now attempt to prove some more results in our (enriched) theory, such as

$$\vdash \forall x \bullet \mathit{grandfather}(x) > x$$

The proof will be written in outline only, but you should convince yourself that you could fill in the missing details, and hence formalise it. To highlight this informality, we will not use our usual layout for the derivation, but will just write down the appropriate sentences surrounded by English. We will also abbreviate our meaningful terms *father* and *grandfather* to f and g. This allows the structure of the proof to be seen more easily, although it might make the semantics less obvious[1]. We start our proof with the axiom

$$\forall x \bullet f(x) > x$$

and instantiate it with the two arbitrary terms a and $f(a)$ giving

$$f(a) > a \qquad \text{and} \qquad f(f(a)) > f(a)$$

Now we can use Axiom 5.

$$(x > y) \wedge (y > z) \Rightarrow (x > z)$$

to give

$$f(f(a)) > a$$

but since we know that $f(f(a)) = g(a)$ by Axiom 3, then

$$g(a) > a$$

[1] This choice between long, meaningful names and short, easily manipulated names presents us with a problem, since we often want both features in one formal system.

and since a was arbitrarily chosen it must be the case that

$$\vdash \forall x \bullet g(x) > x$$

This is an example of an *informal proof*; it has the benefits of being shorter and possibly easier to follow than a totally formal one, and, *if you understand what formal steps are omitted*, then it is adequate for most purposes. Many of the proofs in the rest of this book will be presented informally.

Exercise 4.4 *Find a model for this enriched theory based on the interpretation of Example 4.1.*

Exercise 4.5 *Enrich the theory of the Smiths further to include a predicate for brotherhood, and add an axiom to the effect that men are brothers if, and only if, they have the same father. Use this theory to prove that brothers have the same grandfather, and that any man's father must be older than his brother.*

Exercise 4.6 *Enrich the theory of the Smiths to include a female side of the family, which you must establish. Add the notions of mother, grandmother, sister, wife and husband. Add the idea of blood relationships, and a rule that says no person may marry a blood relation.*

4.2 Uses of Theories

In computer science there are two major applications of formal theory presentations: modelling the real world and specifying artifacts to be built. The distinction between these two uses is subtle, and does not unduly affect the mechanics of the problem, except in how we perceive the task. An inconsistent theory of light, for example, does not mean light does not exist but that the theory is inappropriate[2]. An inconsistent theory acting as a specification for a new database management system, however, means we will have considerable difficulty designing the system.

[2] We might consider it to mean, however, that light *as we perceive it* does not exist.

In this section we are going to restrict our interests to modelling simple systems which we will assume exist, delaying the use of theories for specification until later in the book, when our box of mathematical tools has been extended to allow a more efficient treatment.

A Theory of Boxes

In this section we will capture as a formal theory, using predicate calculus with equality as the underlying logic, a world containing three boxes, called a, b and c, and a table t. Each block can be either standing directly on the table, or stacked on another block. Here are some general properties of this world, expressed using the predicates $above(x, y)$ and $on(x, y)$ to denote x is above, or on, y respectively. First we will say that *above* has the following property

Axiom 1 $\forall x, y, z \bullet (above(x, y) \wedge above(y, z)) \Rightarrow above(x, z)$

Second we will say that if a box is on something then it is above it

Axiom 2 $\forall x, y \bullet on(x, y) \Rightarrow above(x, y)$

We will establish a particular arrangement of our boxes, namely that a and c are on the table, and that b is on top of a

Axiom 3 $on(a, t)$

Axiom 4 $on(c, t)$

Axiom 5 $on(b, a)$

We can now use this theory to reason about the situation in a formal manner. Introducing formality in this way is a possible first step to automating the reasoning process, consequently formal theory presentations of this type are of interest to the Artificial Intelligentsia.

Exercise 4.7 *Prove that b is above the table.*

Exercise 4.8 *Enrich the above theory to include the idea that two objects are at the same level if they are on the same object, and then prove that a and c are on the same level.*

Exercise 4.9 *Enrich the theory further to include the idea that two boxes are the same level if the things they are standing on are themselves at the same level. Add two further boxes, d and e to the world, with d being on the table, and e being on d, then prove that b and e are at the same level.*

Exercise 4.10 *Enrich this theory to include the idea that a box is closed if there is another box on top of it, and open otherwise; then prove that b is open.*

A Theory of a Dastardly Deed

This theory is designed to capture a world in which a crime has been committed. The scenario is as follows

> Only Tom and Harry had keys. Someone stole the money by opening the door. The only way to open the door is with a key.

We will start by identifying the required linguistic components, and giving them an interpretation

$S(x)$ denotes x stole the money

$D(x)$ denotes x opened the door

$K(x)$ denotes x had a key

t denotes Tom

h denotes Harry

We can now capture the above situation formally by producing the appropriate axioms. These are stated without further explanation.

Axiom 1 $\forall x \bullet D(x) \Rightarrow K(x)$

Axiom 2 $\exists x \bullet S(x) \wedge D(x)$

Axiom 3 $\forall x \bullet (K(x) \Rightarrow ((x = t) \vee (x = h)))$

We are now in a position to provide proof that either Tom or Harry stole the money. The formal proof is shown in Figure 4.1

Theorem 4.2 *Prove that*

$$\vdash S(t) \vee S(h)$$

Proof

1	$\exists x \bullet S(x) \wedge D(x)$	axiom 2
2	$S(a) \wedge D(a)$	assumption
3	$S(a)$	2 $\wedge$–elimination
4	$D(a)$	2 $\wedge$–elimination
5	$\forall x \bullet D(x) \Rightarrow K(x)$	axiom 1
6	$D(a) \Rightarrow K(a)$	5 $\forall$–elimination
7	$K(a)$	4, 6 $\Rightarrow$–elimination
8	$\forall x \bullet (K(x) \Rightarrow ((x = t) \vee (x = h)))$	axiom 3
9	$K(a) \Rightarrow ((a = t) \vee (a = h))$	8 $\forall$–elimination
10	$(a = t) \vee (a = h)$	7,9 $\Rightarrow$–elimination
11	$a = t$	assumption
12	$S(t)$	3,11 substitution
13	$S(t) \vee S(h)$	12 $\vee$–introduction
14	$a = h$	assumption
15	$S(h)$	3,14 substitution
16	$S(t) \vee S(h)$	15 $\vee$–introduction
17	$S(t) \vee S(h)$	10–16 $\vee$–elimination
18	$S(a) \wedge D(a) \Rightarrow S(t) \vee S(h)$	2–17 $\Rightarrow$–introduction
19	$\forall x \bullet (S(x) \wedge D(x) \Rightarrow S(t) \vee S(h))$	18 $\forall$–introduction
20	$S(t) \vee S(h)$	1,20 $\exists$–elimination

QED

Figure 4.1: Proof of Guilt

Exercise 4.11 *Enrich this theory to include a statement of Tom's innocence, and hence prove Harry's guilt.*

Exercise 4.12 *Produce a formal theory which has as a model the following situation*

> *All monkeys like bananas. Chimpanzees are monkeys. Basil is a chimpanzee. Monkeys who can press the button get a banana. Monkeys who get what they like are happy. Basil can press the button if the light is on.*

Use this theory to show that either the light is off or Basil is happy.

Exercise 4.13 *Produce a formal theory which has as a model the following situation*

> *There is at most one honest salesperson. Johnson is an honest salesperson. Mary is not Johnson.*

Use this theory to show that Mary is not an honest salesperson.

4.3 The Way Forward

We are now at a stage in the book where we have laid most of our foundations. We have introduced the notion of a formal system, given examples of logics and their associated calculi, and shown how these can be used to present theories. Large constructions, however, seldom proceed at a reasonable pace, even if the foundations are well laid, without some specialist tools. The next portion of this book is devoted to providing a set of tools which can be used within our formal framework.

Consider, for example, the two predicates introduced in this chapter *older than* and *above*. Both of these predicates were concerned with discussing the relations between objects, and they both satisfied an axiom of the form

$$\forall x, y, z \bullet related(x, y) \land related(y, z) \Rightarrow related(x, z)$$

Many relationships satisfy this *transitive* property, and, had we established a theory of transitive relations before we started, we might have been able to use some general results about them.

In the second part of this book we are going to introduce a number of theories which provide us with useful, general results, and upon which other theories can be built. Each theory will be constructed by enriching those previously introduced, adding new descriptive power to the language and new axioms to the deductive apparatus. The starting point for this process is the predicate calculus with equality we have already met, which we will take as our underlying logical system. We do not need to add new inference rules to this logic, as those already introduced will suffice[3].

The method by which these enrichments will be presented is to introduce the objects we wish to formalise, then to introduce any necessary additional notation required, and finally to write as axioms any laws governing the behaviour of these objects. Additional laws governing the objects will then emerge as theorems of the formal system. Where we choose to prove theorems, we have at our disposal all the theorems and derivations previously proved in the book, of course, because they all form part of the formal system being enriched.

Once the techniques of proving theorems in our more powerful formal systems have been established we will feel at liberty to reduce the formality by giving outline proofs only. The reader is encouraged to treat these as additional exercises, filling in the missing detail to provide formal proofs. We will also start to introduce the properties of our objects without explicitly expressing all of them as axioms or theorems, as the formal clutter hinders the speedy development of the subject matter.

As our formal systems increases in descriptive power, so we can start to use them for software engineering purposes. We can, for example, formally specify aspects of a system in terms of the general, mathematical theories introduced by enriching these general theories with axioms constraining the behaviour of the particular system being specified. This means that we will be developing two kinds of theories, those intended as general, mathematical, tools and those intended as specifications of particular systems. In the abstract these are indistinguishable, both being examples of formal systems, but clearly only

[3] Note that any inference rule we might require, such as $\frac{\mathcal{A}}{\mathcal{B}}$, can be presented as an axiom of the form $\mathcal{A} \Rightarrow \mathcal{B}$.

the general theories are intended to be carried forward in subsequent enrichments.

As our theories become more complex, so it is useful to add more structure to their presentation. We will achieve this by introducing a *schema notation* which is particularly well-suited to the task of specification, although it is actually just a general mathematical device.

The general theories we are going to build are those of sets, relations, functions and sequences. These are seen as fundamental to utilising mathematics in software engineering, but they by no means form an exhaustive list of useful bodies of theory. Unfortunately it is impossible in a book of this size to cover all the theories that might prove useful tools, but we hope that you will feel confident enough to spot the need for a new tool when such a need arises, learn to master it, and then use it productively.

It is important when reading the rest of the book, that you bear in mind it is trying to serve like an instruction leaflet for a tool-kit. Each of the tools is worthy of several books in its own right, and as your use of the tools becomes more sophisticated so you may need to refer to the next level of manual. You should also bear in mind that, just as the foundations had to be well laid, tools need to be used frequently in order that you become sufficiently skilled in their use.

Chapter 5

Set Theory

5.1 Sets

The idea of a set is something that is fundamental to modern software engineering, and we hope to convince you of this fact over the next four chapters. Mathematicians have found set theory to be a particularly fruitful area of research, but we shall be interested in it only insofar as it helps us to model software systems and to reason about them.

Intuitively, a *set* is any well-defined collection of objects; what we mean by "well-defined" will be explained as we proceed. The objects in a set are called the *elements* or *members* of the set.

Example 5.1 *The set of all small primes (that is, say, less than 20) is defined as*[1]

$$SmallPrimes \triangleq \{2, 3, 5, 7, 11, 13, 17, 19\}$$

Fortunately, there are only a few small primes, but we impose no restriction on the number of elements that there may be in a set; in particular, it may be infinite. The reader should have no difficulty in imagining what the following (informal) notation means

[1] The notation $N \triangleq E$ means that the N is by definition a name for, and hence equal to, the expression E. N is thus a *syntactic abbreviation* for E, and E and N may readily be interchanged.

Example 5.2 *The set of even natural numbers*

$$EvenNumbers \triangleq \{0, 2, 4, 6, 8, 10, \ldots\}$$

We have adopted the convention that the names of sets begin with a capital letter; we shall attempt to keep to this.

Example 5.3 $IBMproject \triangleq \{ib, jim, steve\}$

As fundamental as the idea of a set is the idea of *set membership.* If x is some object in some set S, we shall write

$$x \in S$$

to describe this fact. We pronounce this "*x belongs to S*" or "*x is in S*". To record the fact that an object x is *not* a member of S, we shall write

$$x \notin S$$

("*x does not belong to S*"). Obviously, $x \notin S$ is simply the negation of $x \in S$

$$x \notin S \Leftrightarrow \neg(x \in S)$$

Example 5.4

$$ib \in IBMproject$$
$$jim \in IBMproject$$
$$mike \notin IBMproject$$

Axiom 5.1 (Set Membership) *A simple fact about set membership is that any x is an element of the singleton set containing it*

$$\forall x \bullet x \in \{x\}$$

Axiom 5.2 (Empty Set) *A set of some interest is the* empty set $\{\}$ *(sometimes also written $\emptyset$). There is no object x in the empty set, and so*

$$\neg \exists x \bullet x \in \{\}$$

Of course, this may also be written as

$$\forall x \bullet \neg(x \in \{\})$$

Example 5.5 *The set of integer solutions to the equation* $x^2 = 5$ *is the empty set.*

Axiom 5.3 (Extension) *The idea of set membership allows us to characterise equality between sets: two sets S and T are* equal *if they both have the same members. That is, if every member of S is also a member of T,* and *if every member of T is also a member of S. If this is so, we write $S = T$, where*

$$(S = T) \Leftrightarrow (\forall x \bullet x \in S \Leftrightarrow x \in T)$$

Reflecting on the axiom of extension allows us to discover two important properties of sets: namely, that there is no concept of order or of repetition of elements.

Example 5.6 *It doesn't matter how we write down the elements of a set if we cannot distinguish between the possible representations using the axiom of extension. Alternative ways of writing two sets that we have already seen are*

$$\begin{aligned} \{ib, ib, ib, jim, jim, steve\} &= IBMproject \\ \{2, 19, 3, 17, 5, 13, 7, 11\} &= SmallPrimes \end{aligned}$$

Some sets are so useful and widely used that we give them special names. The set of *natural numbers* $\mathbb{N}$ is one such set. We shall see how to describe this set rigorously later on, but for now we give a simple intuitive definition for $\mathbb{N}$

$$\mathbb{N} \triangleq \{0, 1, 2, 3, 4, 5, \ldots\}$$

The style of specifying a set by exhibiting all its members is known as specification of a set *in extension*.

We have not said precisely what kinds of things the members of a set can be—we merely said that they were objects of some kind. In fact, sets themselves can be regarded as objects in their own right, so it makes perfect sense to have a set of objects where the objects are sets—*a set of sets*.

Example 5.7 *The set which contains the sets SmallPrimes and EvenNumbers is*

$$\{SmallPrimes, EvenNumbers\}$$
$$= \{\{2,3,5,7,11,13,17,19\},\{0,2,4,6,8,10,\ldots\}\}$$

This set has, of course, just two members.

5.2 Basic Set Theory

In the last section we introduced the fundamental idea of a set. In this section we shall see how to construct new sets from old. Basic set theory is, amongst other things, a language that allows us to do this. We shall see several ways of constructing sets that prove to be so useful that we can use them to define relations, functions, and sequences in later chapters.

Set Comprehension

One of the simplest ways of creating a new set is take an existing set, and to "filter" its members through some predicate. This will have the general form, for some set S and predicate P

$$\{x : S \mid P(x)\}$$

"the set of all x in S that satisfy $P(x)$" We use the notation $x : S$ rather than $x \in S$ to distinguish between the assertion that the name x, introduced at this point, ranges over the values of the set S, and the proposition x belongs to the set S, which may be true or false. As we have already seen, x is a bound variable.

Example 5.8 *The following are all examples of sets described in comprehension*[2]

1. *The set of natural numbers less than 20:* $\{x : \mathbb{N} \mid x < 20\}$.

[2]In these examples, *true* is a shorthand for the predicate *true*(x), which is true in all interpretations; *false* is a shorthand for the predicate *false*(x), which is false in all interpretations.

2. *The set of primary colours other than magenta:*

$$\{x : PrimaryColours \mid x \neq magenta\}$$

where $PrimaryColours \mathrel{\hat{=}} \{cyan, magenta, yellow\}$

3. *The set of prime numbers:*

$$\{x : \mathbb{N} \mid \forall y \bullet (y \in \mathbb{N} \land y \uparrow x) \Rightarrow (y = 1 \lor y = x)\}$$

where $y \uparrow x$ *means "y divides into x exactly".*

4. *The set of all natural numbers:* $\{x : \mathbb{N} \mid true\}$

5. *The empty set:* $\{x : \mathbb{N} \mid false\}$

Exercise 5.1 *Describe the set of small primes (less than 20) in this style.*

The style that we have introduced here is known as specification of a set *in comprehension.* It allows us to formalise the definition of a set succinctly—this is especially important if the set is infinite!

Example 5.9 $\{0, 2, 4, 6, 8, 10, \ldots\} = \{x : \mathbb{N} \mid 2 \uparrow x\}$

Just as the bound variables in a quantified predicate can be renamed at will, so can the bound variables in a set comprehension term.

Example 5.10

$$(\forall x \bullet x \in Even \Rightarrow 2 \uparrow x) \Leftrightarrow (\forall y \bullet y \in Even \Rightarrow 2 \uparrow y)$$

$$\begin{aligned} Even &\mathrel{\hat{=}} \{x : \mathbb{N} \mid 2 \uparrow x\} \\ &= \{y : \mathbb{N} \mid 2 \uparrow y\} \end{aligned}$$

Sometimes it is convenient to specify a set by showing the general pattern of its elements. Any variables used must be bound by saying from which sets they come. This will have the general form, for some set S, and some term $t(x)$

$$\{x : S \bullet t(x)\}$$

The term describes the "shape" of the objects in the set.

Example 5.11 *The following are all subsets of the natural numbers*

1. $\{x : \mathbb{N} \bullet 2x\}$
2. *Squares* $\triangleq \{x : \mathbb{N} \bullet x^2\}$
3. *DoubleSquares* $\triangleq \{x : \mathbb{N} \bullet 2x^2\}$

Exercise 5.2 *Rewrite Squares and DoubleSquares as set comprehension terms of the form* $\{x : X \mid P(x)\}$

This style of specifying a set by giving the pattern for its elements is known as specification of a set *in comprehension by form.*

We can identify two more axioms of our theory of sets which characterise the membership of sets in comprehension

Axiom 5.4

$$x \in \{y : S \mid P(y)\} \Leftrightarrow (x \in S \wedge P(x))$$

Axiom 5.5

$$x \in \{y \in S \bullet t(y)\} \Leftrightarrow (\exists\, y \bullet y \in S \wedge x = t(y))$$

where $t(y)$ *is some term in* y, *such as* y^2 *or* $2y$.

Exercise 5.3 *Establish a connection between the definition of a set in extension and in comprehension by proving the simple theorem*

$$\vdash \forall\, x \bullet x \in S \Rightarrow \{x\} = \{y : S \mid y = x\}$$

We can prove a useful theorem about set comprehension:

Theorem 5.1 *If we replace the predicate in a set comprehension term by a weaker condition, the resulting set may be larger, but contains at least all the elements of the first set*

$$\begin{aligned}\vdash\ & (\forall\, x \bullet P(x) \Rightarrow Q(x)) \Rightarrow \\ & \quad (\forall\, x \bullet x \in \{y : S \mid P(y)\} \Rightarrow x \in \{y : S \mid Q(y)\})\end{aligned}$$

Proof

1	$\forall x \bullet P(x) \Rightarrow Q(x)$	assumption
2	$P(a) \Rightarrow Q(a)$	1 $\forall$-elimination
3	$a \in \{y : S \mid P(y)\}$	assumption
4	$a \in S \wedge P(a)$	3 comprehension
5	$P(a)$	4 $\wedge$-elimination
6	$Q(a)$	2,5 $\Rightarrow$-elimination
7	$a \in S$	4 $\wedge$-elimination
8	$a \in S \wedge Q(a)$	7,6 $\wedge$-introduction
9	$a \in \{y : S \mid Q(y)\}$	8 comprehension
10	$a \in \{y : S \mid P(y)\} \Rightarrow a \in \{y : S \mid Q(y)\}$	3-9 $\Rightarrow$-introduction
11	$\forall x \bullet x \in \{y : S \mid P(y)\} \Rightarrow x \in \{y : S \mid Q(y)\}$	10 $\forall$-introduction
12	$(\forall x \bullet P(x) \Rightarrow Q(x)) \Rightarrow (\forall x \bullet x \in \{y : S \mid P(y)\} \Rightarrow x \in \{y : S \mid Q(y)\}$	1-10 $\Rightarrow$-introduction

QED

Exercise 5.4 *Given the following definitions*

$$MoreThanTen \mathrel{\hat{=}} \{x : \mathbb{N} \mid x > 10\}$$
$$Digits \mathrel{\hat{=}} \{x : \mathbb{N} \mid x \leq 9\}$$
$$Ten \mathrel{\hat{=}} \{10\}$$

prove that

$$MoreThanTen = \{x : \mathbb{N} \mid x \notin Digits \wedge x \notin Ten\}$$

Specification in Comprehension

The full form of an implicit set definition given in comprehension is a combination of the two forms that we have already met

$$\{x : S \mid P(x) \bullet t(x)\}$$

Here x is a bound variable which satisfies $P(x)$; the elements of the generated set all have the form $t(x)$.

Example 5.12 *The set of all numbers which are one more than an even cube is described by*

$$\{x : \mathbb{N} \mid x^3 \in \mathit{Even} \bullet x + 1\} = \{1, 9, 65, \ldots\}$$

The following identities hold

$$\{x : S \mid P(x)\} = \{x : S \mid P(x) \bullet x\}$$
$$\{x : S \bullet t(x)\} = \{x : S \mid \mathit{true} \bullet t(x)\}$$

5.3 Power Sets

The set

$$\mathit{SmallNat} = \{x : \mathbb{N} \mid x < 20\}$$

is a *subset* of $\mathbb{N}$ in the sense that every member of *SmallNat* is also a member of $\mathbb{N}$.

Exercise 5.5 *Prove that this is indeed the case by showing that*

$$\forall x \bullet (x \in \mathit{SmallNat} \Rightarrow x \in \mathbb{N})$$

Of course, there are other subsets of $\mathbb{N}$, rather a lot in fact. We can use set extension and set comprehension to describe many of them, but if we want to describe them all it will take us from here to eternity. So, we would like to have some succinct way of representing the set of all subsets of a set S. We introduce the *power set* of S, written variously in the literature as

$$\mathbb{P}\, S, \mathcal{P} S, \mathit{or}\ 2^S$$

Example 5.13

$$\mathbb{P}\{a, b, c\} = \{\{\}, \{a\}, \{b\}, \{c\}, \{a, b\}, \{b, c\}, \{a, c\}, \{a, b, c\}\}$$

Notice that, for any set S, both $\{\}$ and S are subsets of S. In general,

Axiom 5.6 (Power Set)

$$S \in \mathbb{P}\, T \Leftrightarrow (\forall x \bullet x \in S \Rightarrow x \in T)$$

This is an axiom which characterises the power set.

Exercise 5.6 *Write out in full the power sets of*

1. $\{0, 1\}$
2. $\{5\}$
3. $\{\}$
4. $\{x : \mathbb{N} \mid 2 \uparrow x \wedge x \leq 6\}$
5. $\{x : Primes \mid x \neq 2 \wedge 2 \uparrow x\}$

If a set S is finite and contains precisely n elements, then the power set of S, $\mathbb{P}\, S$, will have 2^n elements. That is why the power set of S is sometimes written as 2^S.

5.4 Cartesian Products

Just as set comprehension is a powerful way of describing a particular subset of a set, so the power set is a powerful way of describing all the subsets of a set. In this section we look at a different way of forming new sets from old by using ordered pairs of objects.

Example 5.14 *A square on a chess board is an ordered pair (v, h), where v is the distance in squares from the bottom of the board, and h is the distance in squares from the left-hand edge.*

Let (x, y) be the ordered pair of objects x and y, where x is the *first* object, and y is the *second* object. Two ordered pairs are equal just when they have equal first and equal second components

$$(x_1, y_1) = (x_2, y_2) \Leftrightarrow (x_1 = x_2 \wedge y_1 = y_2)$$

We can treat ordered pairs as objects, and think about sets of ordered pairs. If S and T are sets, then the set of *all* ordered pairs (x, y), where

$$x \in S \wedge y \in T$$

is called the *cartesian product*[3] of S and T. The cartesian product of S and T is written

$$S \times T$$

which we pronounce "*S cross T*", and which consists of all the ordered pairs (x, y) where $x \in S$ and $y \in T$. So, we can express this set in comprehension by form

$$S \times T = \{x : S; y : T \bullet (x, y)\}$$

Example 5.15 *The following are all examples of cartesian products*

1. *ChessBoard* $\triangleq \mathbb{N}_8 \times \mathbb{N}_8$ *where*

$$\mathbb{N}_8 \triangleq \{x : \mathbb{N} \mid 1 \leq x \leq 8\}$$

So a chess board can be represented by the set of pairs

$$\mathbb{N}_8 \times \mathbb{N}_8 = \{x, y : \mathbb{N} \mid 1 \leq x \leq 8 \wedge 1 \leq y \leq 8 \bullet (x, y)\}$$

2. *The "Cartesian Plane"*

$$CarteBlanche \triangleq \mathbb{N} \times \mathbb{N}$$

We can imagine CarteBlanche representing an infinite piece of graph paper, with a fixed margin at the bottom and at the left, say.

3. *PaintingByNumbers* $= \mathbb{N} \times$ *PrimaryColours*

Exercise 5.7 *Write out in full the cartesian products*

1. $\{0,1\} \times \{0,1\}$
2. *PrimaryColours* $\times$ *PrimaryColours*
3. *IBMproject* $\times \{\}$
4. $\{\} \times \{\}$
5. $\{1,2\} \times \{a\}$

---

[3] Named after René Descartes, the French mathematician.

6. $\{\{\}\} \times \{a\}$

7. $\{a, b\} \times \mathbb{P}\{a, b\}$

8. $(\{a\} \times \{b, c\}) \times \{c\}$

As we did with power sets, so we can characterise the membership of cartesian products

Axiom 5.7 (Cartesian Product)

$$x \in (S \times T) \Leftrightarrow \exists\, y, z \bullet (y \in S \wedge z \in T \wedge x = (y, z))$$

We can generalise our notion of the cartesian product of two sets to that of many sets:

Example 5.16 *Let $S = \{a, b\}$, then*

$$\begin{array}{l} S \times S \times S = \\ \quad \{(a,a,a),(a,a,b),(a,b,a),(b,a,a), \\ \quad\quad (a,b,b),(b,a,b),(b,b,a),(b,b,b)\} \end{array}$$

In this example, $S \times S \times S$ is a set of *triples.* Notice that this is different from $S \times (S \times S)$, a set of pairs, the second component of which is itself a pair, and $(S \times S) \times S$, a set of pairs, the first component of which is itself a pair.

Exercise 5.8 *Write out in full the sets $S \times (S \times S)$ and $(S \times S) \times S$.*

Exercise 5.9 *Describe the set of degrees of arc, Degrees, as a subset of the real numbers $\mathbb{R}$; and then describe the set Clock as a subset of Degrees $\times$ Degrees, where the first component is the hour hand and the second is the minute hand. Make sure that there is an appropriate relationship between the hands.*

Exercise 5.10 *Describe the set of inverted clocks Inverted, where the clock has been rotated so that 6 o'clock is at the top, and 12 o'clock is at the bottom.*

Exercise 5.11 *How many elements are there in Inverted?*

5.5 Predicates and Sets

In the last chapter we encountered quantified predicates such as

$$\exists x \bullet x > 4$$
$$\exists x \bullet x = \text{"}a\text{"}$$
$$\forall x \bullet x \neq mike$$
$$\forall x \bullet (\exists y \bullet y > x)$$

We can now start to give a more rigorous meaning to these expressions by saying over which sets the bound variables range. It would be more useful to write the following

Example 5.17

$$\exists x \bullet x \in \mathsf{N} \wedge x > 4$$
$$\exists x \bullet x \in Char \wedge x = \text{"}a\text{"}$$
$$\forall x \bullet x \in IBMproject \Rightarrow x \neq mike$$
$$\forall x \bullet x \in \mathsf{N} \Rightarrow (\exists y \bullet x \in \mathsf{N} \wedge y > x)$$

In the case of the existentially quantified formulæ, we must find an x that satisfies the predicate and *belongs to the set. In the case of the universally quantified formulæ, only those values of x drawn from the set need satisfy the predicate. Hence the conjunctions and implications in the above formulæ.*

The usefulness of being more precise about the bound variables is in danger of becoming outweighed by the increase in formal clutter, so we introduce the following shorthands

$$\exists x : X \bullet P(x) \;\; \textit{for} \;\; \exists x \bullet x \in X \wedge P(x)$$
$$\forall x : X \bullet P(x) \;\; \textit{for} \;\; \forall x \bullet x \in X \Rightarrow P(x)$$

Thus we can rewrite our examples

$$\exists x : \mathsf{N} \bullet x > 4$$
$$\exists x : Char \bullet x = \text{"}a\text{"}$$
$$\forall x : IBMproject \bullet x \neq mike$$
$$\forall x : \mathsf{N} \bullet (\exists y : \mathsf{N} \bullet y > x)$$

Notice that universally quantifying over the empty set is a tautology, which we can show by the following theorem

Theorem 5.2

$$\vdash \forall x : \{\} \bullet P(x)$$

Proof

1	$\forall x \bullet \neg(x \in \{\})$	Empty set axiom
2	$\neg(a \in \{\})$	1 $\forall$-elimination
3	$a \in \{\} \Rightarrow P(a)$	2 vac $\Rightarrow$-introduction
4	$\forall x \bullet x \in \{\} \Rightarrow P(x)$	3 $\forall$-introduction
5	$\forall x : \{\} \bullet P(x)$	5 by definition

QED

Exercise 5.12 *Prove that existentially quantifying over the empty set gives us a contradiction by proving the theorem*

$$\vdash \neg \exists x : \{\} \bullet P(x)$$

5.6 Types

When people use set theory to help to specify software systems, they often include some notion of types. For example, when we introduced bound variables in set definitions, we were careful to say precisely over which sets their values may range; we have just started to be more rigorous about quantified predicates. In this sense, we say that we are working in a *typed set theory*, and we call $\mathbb{N}$, $\mathbb{P}(B \times C)$, and *SmallPrimes* in the following formulæ *types*:

$$\forall x : \mathbb{N} \bullet P(x)$$
$$\exists R : \mathbb{P}(B \times C) \bullet R = \{\}$$
$$\exists p : \mathit{SmallPrimes} \bullet 15 < p < 20$$

The notion of type that we shall stick to is a simple one: a type is a *maximal* set, in the sense that values may belong to just one type. Thus, for the moment we shall have nothing to do with the complications of subtypes or such notions. So, if we really do regard *SmallPrimes* as a type, then we cannot also have $\mathbb{N}$ as a type in the same system, otherwise 3 would belong to two different types. However, if T is a type, and S is a subset of T, we often write

$$\exists\, x : S \bullet P(x)$$

as a shorthand for

$$\exists\, x : T \bullet x \in S \wedge P(x)$$

The reader will notice that such shorthands may be used whenever a variable is introduced in our notation

$$\{x : S \bullet t(x)\} = \{x : T \mid x \in S \bullet t(x)\}$$
$$(\forall\, x : S \bullet P(x)) \Leftrightarrow (\forall\, x : T \bullet x \in S \Rightarrow P(x))$$

We use types to avoid certain unpleasant foundational problems that might creep into our mathematics otherwise. The well-known example of this is Russell's paradox. Let

$$S = \{X \mid X \notin X\}$$

We have met sets of sets before, so it shouldn't seem unreasonable to encounter a set of sets, none of whose elements belong to themselves (except, of course, that you know that you are being set up!). Now, is $S \in S$? If it is,

1	$S \in S$	assumption
2	$S \in \{X \mid X \notin X\}$	1 S-defn
3	$S \notin S$	2 Comprehension

which is clearly a contradiction. So, we were wrong, it must be that $S \notin S$

1	$S \notin S$	assumption
2	$\neg(S \in S)$	1 $\notin$-defn
3	$\neg(S \in \{X \mid X \notin X\})$	2 S-defn
4	$\neg(S \notin S)$	3 Comprehension
5	$\neg\neg(S \in S)$	4 $\notin$-defn
6	$S \in S$	5 $\neg$-elimination

which is also a contradiction. We are stumped both ways: a *paradox*. We could not have written the definition of S if we had been more careful about the types of its elements, could we?

Exercise 5.13 *Try writing a typed definition of S; exactly what is the problem?*

5.7 Summary

In this short chapter we have laid the foundations of elementary set theory. We introduced the primitive notions of set and set membership, upon which the theory rests. We have seen three important ways of specifying a set: in extension, in comprehension, and in comprehension by form. Starting from these basics, more complicated sets can be constructed, using the power set and cartesian product constructors that we have introduced. The final section introduced the simple type system that we shall use throughout the rest of the book.

Chapter 6

Relations

6.1 An Introduction to Relations

Binary Relations on a Set

In Chapter 3 we introduced a way of describing predicates on objects, and in Chapter 5 we introduced a way of describing the sets to which these objects may belong. In this section we show how to put these ideas together to describe relationships between objects taken from particular sets.

Example 6.1 *In the last chapter we informally introduced a relation between natural numbers* $x \uparrow y$ *to mean* x *divides into* y *exactly . We can formalise this*

$$\forall x, y : \mathbb{N} \bullet (x \uparrow y \Leftrightarrow \exists z : \mathbb{N} \bullet x \times z = y)$$

That is, y *is some exact multiple of* x.

But we haven't quite completed the definition of this relation. We would like to make it explicit that it is a *binary relation on the natural numbers*[1]. We call the set of all possible binary relations on $\mathbb{N}$

$$\mathbb{N} \leftrightarrow \mathbb{N}$$

[1] If a relation is between members drawn from the same set X, then we say that it is a relation *on* X, and that it is *homogeneous*.

Then it is clear that $\uparrow \in \mathbb{N} \leftrightarrow \mathbb{N}$, if we think of $\uparrow$ as just another object. At the moment we can think of $\leftrightarrow$ as a kind of set constructor, just like $\times$ or $\mathbb{P}$. Our formal definition of $\uparrow$ now says that it is a binary relation on the naturals, and then describes the property that $\uparrow$ must satisfy. We will write this as

$$\begin{array}{|l}
_ \uparrow _ : \mathbb{N} \leftrightarrow \mathbb{N} \\
\hline
\forall x, y : \mathbb{N} \bullet \\
\qquad x \uparrow y \Leftrightarrow \exists z : \mathbb{N} \bullet x \times z = y
\end{array}$$

This notation

$$\begin{array}{|l}
x : S \\
\hline
P(x)
\end{array}$$

introduces a mathematical variable x whose values are drawn from the set S, and which must satisfy the predicate $P(x)$. The symbol $\uparrow$ simply stands for a mathematical variable, just like x, y, or z. The underscores in the definition of $\uparrow$ say that it is an *infix* relation; they tell us where to put the arguments. Symbols are often described as *infix*, such as addition ($\_ + \_$) or subtraction ($\_ - \_$); *prefix*, such as negation ($\neg$) or the power set constructor ($\mathbb{P}$); *postfix*, such as factorial ($\_ !$) or inverse ($\_ ^{-1}$); *outfix*, such as determinant of a matrix ($\| \_ \|$) or the semantic evaluation function in denotational semantics ($[\![\_]\!]$); or *distfix*, such as the programming language alternative construct

if _ then _ else _ fi

or the integral of a function

$$\int_{\_}^{\_} \_ \, dx$$

We use the underscores to say which kind of operator we are dealing with: infix, prefix, postfix, outfix, or distfix. By default, we expect a symbol to be *infix*.

Thus we can say $2 \uparrow 6$, which is true, and $3 \uparrow 7$, which is false.

Exercise 6.1 *Describe a relation*

$$\_overlap\_ : (\mathbb{P}\, X) \leftrightarrow (\mathbb{P}\, X)$$

that holds between two sets when they "overlap", that is, have at least one member in common.

To explore the idea of relations further, we can think of more complicated examples.

State Transitions as Relations

Consider a very simple lift lobby in a building, with just a single button, a single light, and a single pair of doors. Define the following sets, each containing two distinct elements

$$B = \{pressed, released\}$$
$$L = \{lit, unlit\}$$
$$D = \{opened, closed\}$$

A button can be pressed or not pressed, a light can be lit or unlit, and the doors can be opened or closed. The *state* of the lift lobby at any instant is a triple (b, l, d) drawn from

$$LobbyState \;\hat{=}\; B \times L \times D$$

Various events can happen in the lift lobby to change this state: *press*, *release*, *on*, *off*, *open*, and *close*. We shall describe some of these as *relations* between states of the lobby. We can press the button whenever it is released

$$\_press\_ : LobbyState \leftrightarrow LobbyState$$

$$\begin{array}{l} \forall\, b, b' : B;\ l, l' : L;\ d, d' : D \bullet \\ \quad (b, l, d)\ press\ (b', l', d') \Leftrightarrow \\ \qquad b = released \land b' = pressed \land l = l' \land d = d' \end{array}$$

The transition *press* may occur if the button is *released*; in the new state, the button becomes *pressed*. The state of the light and of the door remain unaffected by *press*ing. The definition describes how two states are related by the transition *press*. The state before the transition has been represented by (b, l, d), and the state after by (b', l', d'). We could have used a wide variety of notations, but this convention seems convenient.

Having said that *press* is a relation between lobby states the universal quantification adds nothing, but just clutters our definition. In future we shall often omit the universal quantification and consider the apparently free variables to be implicitly universally quantified. This is standard practice in mathematics and engineering.

The light can go on only if the button is pressed and the doors are closed

$$
\begin{array}{|l}
\_on\_ : LobbyState \leftrightarrow LobbyState \\
\hline
(b, l, d)\ on\ (b', l', d') \Leftrightarrow \\
\qquad b = pressed \wedge b' = pressed \wedge \\
\qquad l = unlit \wedge l' = lit \wedge \\
\qquad d = closed \wedge d' = closed
\end{array}
$$

The door can open only if it is closed and either the button is pressed, or the light is on

$$
\begin{array}{|l}
\_open\_ : LobbyState \leftrightarrow LobbyState \\
\hline
(b, l, d)\ open\ (b', l', d') \Leftrightarrow \\
\qquad (b = pressed \vee l = lit) \wedge \\
\qquad b = b' \wedge l = l' \wedge \\
\qquad d = closed \wedge d' = opened
\end{array}
$$

Exercise 6.2 *Write similar definitions for release, off, and close.*

Maplets

If two objects $x \in S$ and $y \in T$ are related by a relation $R \in S \leftrightarrow T$, we shall write

$$x \mapsto y$$

"*x is related to y*", or "*x maps to y*". $x \mapsto y$ is called a "maplet" and is an element of the relation R

$$x \mapsto y \in R$$

For example,

$$(released, lit, opened) \mapsto (pressed, lit, opened) \in press$$

Using this notation, we can write relations out in full (if they are small enough!) as sets of maplets.

Example 6.2

$$\begin{aligned}
press = \{&(released, lit, opened) \mapsto (pressed, lit, opened),\\
&(released, lit, closed) \mapsto (pressed, lit, closed),\\
&(released, unlit, opened) \mapsto (pressed, unlit, opened),\\
&(released, unlit, closed) \mapsto (pressed, unlit, closed)\}\\
on = \{&(pressed, unlit, closed) \mapsto (pressed, lit, closed)\}\\
open = \{&(pressed, unlit, closed) \mapsto (pressed, unlit, opened),\\
&(released, lit, closed) \mapsto (released, lit, opened),\\
&(pressed, lit, closed) \mapsto (pressed, lit, opened)\}
\end{aligned}$$

Exercise 6.3 *Write out in full the relations release, off, and close.*

The Empty Relation

Since there is nothing to stop us writing a predicate that is unsatisfiable, we must consider the idea of an *empty relation:* a relation that relates no objects. Since there are no maplets in the empty relation, we can write it as the empty set of maplets, $\{\}$.

Example 6.3 *Suppose that we have the following binary relation that holds when two numbers are strictly greater than each other*

$$\begin{array}{|l}
_ \bowtie _ : \mathbb{N} \leftrightarrow \mathbb{N} \\
\hline
m \bowtie n \Leftrightarrow ((n > m) \wedge (m > n))
\end{array}$$

This is a definition that cannot be satisfied by any pair of natural numbers, so clearly

$$\neg \exists\, x, y : \mathbb{N} \bullet x \bowtie y$$

or, equivalently,

$$\forall x, y : \mathsf{N} \bullet (x \mapsto y) \notin \bowtie$$

We can write the definition of the relation $\bowtie$ *in extension as*

$$\begin{array}{|l} \_\bowtie\_ : \mathsf{N} \leftrightarrow \mathsf{N} \\ \hline \bowtie = \{\} \end{array}$$

The relation $\bowtie$ *is an empty relation.*

The Full Relation and A Model for Relations

The representation for writing relations that we have adopted has suggested a way of modelling them: as sets of pairs. We have chosen to write these pairs as $x \mapsto y$ to emphasise the relational nature, but we could just as well have written the pairs in the notation for objects drawn from cartesian products: (x, y). So, the notation $x \mapsto y$ is just a different way of writing the ordered pair (x, y).

The *full relation* relates every possible pair of elements, and is therefore simply the cartesian product of the source and the target of the relation.

Example 6.4 *Let* $A \mathrel{\widehat{=}} \{1,2\}$ *and* $B \mathrel{\widehat{=}} \{potato, tomato\}$, *then*

$$R = \{1 \mapsto potato, 2 \mapsto potato, 1 \mapsto tomato, 2 \mapsto tomato\}$$

is the full relation between A *and* B; *thus we have*

$$R = A \times B$$

Obviously a relation

$$\_R\_ : A \leftrightarrow B$$

isn't simply $A \times B$, since that would include *all possible* pairs from A and B, and the relation may be more selective than that. Instead, R is a *subset* of $A \times B$. This allows for it to be *empty*, *full* ($A \times B$ itself), or anywhere in between. Thus, we have chosen that

$$A \leftrightarrow B = \mathbb{P}(A \times B)$$

that is, the set of all possible relations between members of A and B, namely $A \leftrightarrow B$, is the *power set* of $A \times B$. It may perhaps be regarded as unfortunate when a symmetric shape is chosen to represent an idea which is not symmetric, for instance, for distinct S and T

$$S \leftrightarrow T \neq T \leftrightarrow S$$

because

$$\mathsf{P}(S \times T) \neq \mathsf{P}(T \times S)$$

However, we are not the first people to make such a choice, for example "†" for map overwrite in VDM, or even subtraction in arithmetic!

Example 6.5 *The divides relation*

$$\_ \uparrow \_ : \mathbb{N} \leftrightarrow \mathbb{N}$$

is a set of ordered pairs of $\mathbb{N}$. *That is,*

$$\uparrow \in \mathsf{P}(\mathbb{N} \times \mathbb{N})$$

This justifies our thinking of $\uparrow$, *press, on, and open simply as objects, that is, things that can be members of sets.*

Exercise 6.4 *Describe* $\uparrow$ *using a set comprehension term.*

Heterogeneous Relations

The examples of relations that we have described so far have all been *homogeneous* in the sense that they are binary relations on some set ($\mathbb{N}$ for $\uparrow$ and *LobbyState* for the lift lobby operations). Some relations relate different types of objects; these relations are called *heterogeneous relations*.

Example 6.6 *The following are heterogeneous relations:*

1. *Define the set of people who work on the IBM project in Oxford, and the set of really nice malt whiskies to be*

$$\begin{aligned} IBMproject &\triangleq \{ib, jim, steve\} \\ IslayMalt &\triangleq \{ardbeg, caolila, laphroaig\} \end{aligned}$$

then we formalise a particular set of preferences amongst the project members

$$\begin{aligned} & Drinks \mathrel{\widehat=} \\ & \quad \{ib \mapsto ardbeg, jim \mapsto caolila, ib \mapsto caolila, \\ & \quad ib \mapsto laphroaig, steve \mapsto laphroaig\} \end{aligned}$$

2. *Define the set of interesting subjects that the people who work on the IBM project teach to be*

$$Subjects \mathrel{\widehat=} \{csp, refinement, z, pascal\}$$

then if everyone teaches everything

$$Teaches \mathrel{\widehat=} IBMproject \times Subjects$$

3. *Define the relation that holds between a number and a pair of numbers just when the number is the maximum of the pair*

$$\_pmax\_ : \mathbb{N} \leftrightarrow (\mathbb{N} \times \mathbb{N})$$
$$x\ pmax\ (y, z) \Leftrightarrow (y \geq z \land x = y) \lor (z \geq y \land x = z)$$

4. *Define the relation that holds between a number and a set of numbers just when the number is the maximum of the set*

$$\_smax\_ : \mathbb{N} \leftrightarrow \mathbb{P}\,\mathbb{N}$$
$$x\ smax\ Y \Leftrightarrow (\forall y : Y \bullet x \geq y) \land x \in Y$$

Exercise 6.5 *Prove the following*[2]

1. $13\ pmax\ (7, 13)$
2. $8\ smax\ \{0, 2, 4, 6, 8\}$
3. $\forall x, y, z : \mathbb{N} \bullet x\ pmax\ (y, z) \Leftrightarrow x\ smax\ \{y, z\}$

---

[2] You might find the following lemma useful

$$(a = b \lor a = c) \Leftrightarrow a \in \{b, c\}$$

The Inverse of a Relation

We can define the inverse of a relation $\_R\_ : A \leftrightarrow B$ as follows in comprehension by form[3]

$$R^{-1} \mathrel{\widehat{=}} \{x : A; y : B \mid x\ R\ y \bullet y \mapsto x\}$$

It contains all those pairs in R, but with their order inverted.

Example 6.7 *Given a relation parentof which describes the relationship between parents and their children, we can construct the inverse relation* $parentof^{-1}$ *which describes the relationship between children and their parents. We might call the relation* $parentof^{-1}$ *by the name childof.*

Example 6.8 *The relation Drinks was introduced in Example 6.6; we can define the relation*

$$Drunkenby \mathrel{\widehat{=}} Drinks^{-1}$$

where

$$Drinks^{-1} = \\ \{ardbeg \mapsto ib, caolila \mapsto jim, caolila \mapsto ib, \\ laphroaig \mapsto ib, laphroaig \mapsto steve\}$$

Exercise 6.6 *Devise a set of useful properties of inversion.*

The Identity Relation

The *identity relation* on a set S relates each member of S to itself

$$Id_S \mathrel{\widehat{=}} \{x : S \bullet x \mapsto x\}$$

Example 6.9 *The identity on Islay malt whiskies is*

$$Id_{IslayMalt} \\ = \{ardbeg \mapsto ardbeg, caolila \mapsto caolila, laphroaig \mapsto laphroaig\}$$

[3] This is only a working definition of inversion, enough to give us some intuition. In Chapter 7 we shall give a formal definition.

The Subset Relation

We can define our notion of subset as a relation between sets: a set S is a *subset* of another set T if every member of S is also a member of T. Formally, for some set X

$$
\begin{array}{l}
[X] \\
\hline
\_ \subseteq \_ : \mathbb{P}\, X \leftrightarrow \mathbb{P}\, X \\
\hline
\forall S, T : \mathbb{P}\, X \bullet \\
\quad S \subseteq T \Leftrightarrow \forall s : X \bullet s \in S \Rightarrow s \in T \\
\hline
\end{array}
$$

This notation

$$
\begin{array}{l}
[X] \\
\hline
v : X \\
\hline
P(v) \\
\hline
\end{array}
$$

is used to introduce the mathematical variable v of type X, where the particular set X is a parameter to the definition. It is called a *generic definition.* Instead of describing a subset relation for every possible type, we have been economical, and written just a single definition.

The *proper* subset relation is one that does not include the possibility of equality

$$
\begin{array}{l}
[X] \\
\hline
\_ \subset \_ : \mathbb{P}\, X \leftrightarrow \mathbb{P}\, X \\
\hline
\forall S, T : \mathbb{P}\, X \bullet \\
\quad S \subset T \Leftrightarrow S \subseteq T \land S \neq T \\
\hline
\end{array}
$$

Example 6.10 *The following are all examples of the subset relations*

1. $\{\} \subseteq \{1, 2, 3\}$
2. $\{1, 2, 3\} \subseteq \{1, 2, 3\}$
3. $\{1, 2\} \subset \{1, 2, 3\}$
4. $\{\} \subset \{1, 2, 3\}$

Source and Target, Domain and Range

If R is a relation between A and B, we call A the *source* of the relation and B the *target*. Not every member of A need be related to some member of B by R; those that are are said to belong to the *domain* of R

$$dom\ R \ \widehat{=}\ \{a : A \mid \exists\, b : B \bullet a\ R\ b\}$$

for $R \in A \leftrightarrow B$. The *range* of R is the set of all elements of B to which R relates something taken from its domain

$$ran\ R \ \widehat{=}\ \{b : B \mid \exists\, a : A \bullet a\ R\ b\}$$

for $R : A \leftrightarrow B$. Clearly, for $R \in A \leftrightarrow B$

$$dom\ R \subseteq A \wedge ran\ R \subseteq B$$

6.2 Basic Theory of Relations

In this section we introduce some basic notions in the theory of relations.

Relational Composition

Consider the following set of vertices

$$V = \{A, B, C, D\}$$

Let *Tet* be the relationship between the vertices of a tetrahedron which relates each vertex to every other

$$Tet = \{x, y : V \mid x \neq y \bullet x \mapsto y\}$$

We can draw this as a picture as in Figure 6.1.

If we start at A and travel to C, we could traverse the side $A \mapsto C$, or use an intermediate vertex B or D. That is, we could have traversed from A to B $(A \mapsto B)$ and then B to C $(B \mapsto C)$; or

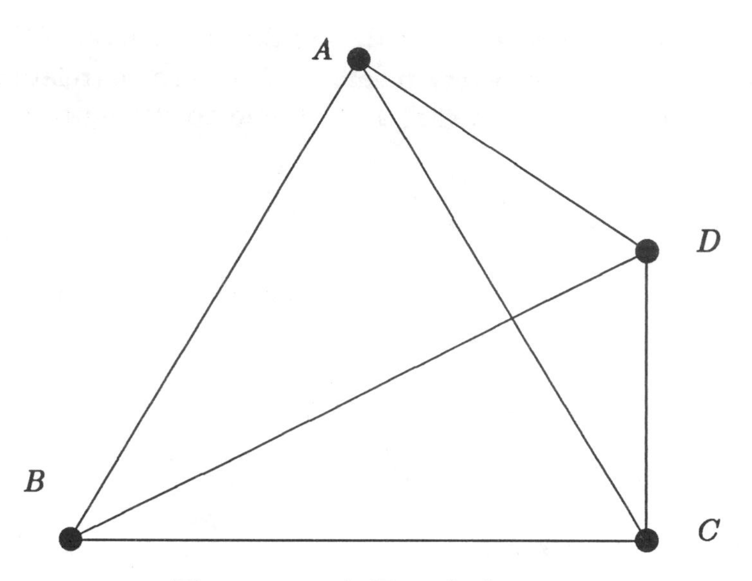

Figure 6.1: A Tetrahedron

from A to D (since $A \mapsto D$) and then D to C (since $D \mapsto C$). We still end up in the same place: C. Let's define a new relation Tet^2 that relates two vertices in the tetrahedron *via* an intermediate vertex

$$Tet^2 = \\ \{x, z : V \mid (\exists\, y : V \bullet x \mapsto y \in Tet \wedge y \mapsto z \in Tet) \bullet x \mapsto z\}$$

Exercise 6.7 *Show that*

$$\begin{aligned} Tet^2 &= V \times V \\ &= \{x, y : V \mid x \mapsto y \in Id_V \vee x \mapsto y \in Tet \bullet x \mapsto y\} \end{aligned}$$

Exercise 6.8 *Another way of getting from A to C is to use two intermediate vertices; list all such routes.*

This relation between two vertices which are connected *via* two intermediate vertices can be defined

$$Tet^3 = \\ \{w, z : V \mid (\exists\, x, y \in V \bullet \{w \mapsto x, x \mapsto y, y \mapsto z\} \subseteq Tet) \bullet w \mapsto z\}$$

Tet^3 because it uses *three* sides.

Exercise 6.9 *Show that*

$$\begin{array}{l} Tet^3 = \\ \{x, y : V \mid (x \mapsto y \in Id_V) \lor (x \mapsto y \in Tet) \lor (x \mapsto y \in Tet^2) \bullet \\ \quad x \mapsto y\} \end{array}$$

The fact that we can travel from A to C *via* some intermediate vertex B (that is, that $A \mapsto C \in Tet^2$) is due to the two facts that $A \mapsto B \in Tet$ *and* $B \mapsto C \in Tet$. Here we have *composed* the relation with itself. We can formalise this: let $R \mathbin{;} S$ be the *relational composition* of R with S, then

$$R \mathbin{;} S \triangleq \{x : A; z : C \mid (\exists\, y : B \bullet x\ R\ y \land y\ S\ z) \bullet x \mapsto z\}$$

for $R : A \leftrightarrow B, S : B \leftrightarrow C$. That is,

$$a\ (R \mathbin{;} S)\ c$$

if there is some b such that $a\ R\ b$ and $b\ S\ c$. Thus,

$$\begin{array}{l} Tet \mathbin{;} Tet \\ \quad = \{x, z : V \mid (\exists\, y : V \bullet \{x \mapsto y, y \mapsto z\} \subseteq Tet) \bullet x \mapsto z\} \\ \quad = Tet^2 \end{array}$$

$$\begin{array}{l} (Tet \mathbin{;} Tet) \mathbin{;} Tet \\ = \{w, z : V \mid (\exists\, y : V \bullet w \mapsto y \in (Tet \mathbin{;} Tet) \land y \mapsto z \in Tet) \bullet \\ \quad w \mapsto z\} \\ = \{w, z : V \mid (\exists\, y : V \bullet \exists\, x : V \bullet \\ \quad w \mapsto x \in Tet \land x \mapsto y \in Tet \land y \mapsto z \in Tet)\} \\ = \{w, z : V \mid (\exists\, y, x : V \bullet \\ \quad w \mapsto x \in Tet \land x \mapsto y \in Tet \land y \mapsto z \in Tet)\} \\ = Tet^3 \end{array}$$

Exercise 6.10 *Prove that* $\mathbin{;}$ *is associative, that is*

$$(R \mathbin{;} S) \mathbin{;} T = R \mathbin{;} (S \mathbin{;} T)$$

6.3 Special Relations

Homogeneous relations can be classified in various ways. Let

$$\_R\_ : S \leftrightarrow S$$

for some set S. These properties can be expressed formally

1. R is *reflexive* if it contains the identity relation on S: it relates every member of S to itself.
   R is reflexive $\Leftrightarrow \forall x : S \bullet x\ R\ x$

2. It is *irreflexive* if it relates no member of S to itself.
   R is irreflexive $\Leftrightarrow \neg \exists x : S \bullet x\ R\ x$.

3. R is *symmetric* if whenever $x\ R\ y$, we also have $y\ R\ x$.
   R is symmetric $\Leftrightarrow \forall x, y : S \bullet x\ R\ y \Rightarrow y\ R\ x$.

4. R is *antisymmetric* if whenever $x\ R\ y$ and $y\ R\ x$, we have that x and y are the same.
   R is antisymmetric $\Leftrightarrow \forall x, y : S \bullet x\ R\ y \wedge y\ R\ x \Rightarrow x = y$.

5. It is *asymmetric* if whenever $x\ R\ y$ we cannot also have $y\ R\ x$.
   R is asymmetric $\Leftrightarrow \forall x, y : S \bullet \neg(x\ R\ y \wedge y\ R\ x)$

6. R is *transitive* if whenever $x\ R\ y$ and $y\ R\ z$ we also have $x\ R\ z$.
   R is transitive $\Leftrightarrow \forall x, y, z : S \bullet x\ R\ y \wedge y\ R\ z \Rightarrow x\ R\ z$.

7. R is an *equivalence relation* if it is reflexive, symmetric, and transitive.

We can use the definitions to prove properties of some of the relations that we have introduced in this chapter.

Example 6.11 *Subset is a reflexive relation: every set is a subset of itself. This law mirrors the law that says that every set is a member of its own power set.*

$$\forall S : \mathbb{P}\, X \bullet S \subseteq S$$

Proof: For arbitrary $a : X$ and $B : \mathbb{P}\, X$

1	$a \in B \Rightarrow a \in B$	TI($\vdash P \Rightarrow P$)
2	$\forall x : X \bullet x \in B \Rightarrow x \in B$	1 $\forall$-introduction
3	$B \subseteq B$	2 $\subseteq$-definition
4	$\forall S : \mathbb{P}\, X \bullet S \subseteq S$	3 $\forall$-introduction

QED

Example 6.12 *Subset is an antisymmetric relation; in fact this law is sometimes used to prove the equality of two sets*

$$\forall S, T : \mathbb{P}\, X \bullet S \subseteq T \wedge T \subseteq S \Rightarrow S = T$$

Proof: For arbitrary $A, B : \mathbb{P}\, X$

1	$A \subseteq B \wedge B \subseteq A$	assumption
2	$A \subseteq B$	1 $\wedge$-elimination
3	$\forall x : X \bullet x \in A \Rightarrow x \in B$	2 $\subseteq$-defn
4	$c \in A \Rightarrow c \in B$	3 $\forall$-elimination
5	$B \subseteq A$	1 $\wedge$-elimination
6	$\forall x : X \bullet x \in B \Rightarrow x \in A$	5 $\subseteq$-defn
7	$c \in B \Rightarrow c \in A$	3 $\forall$-elimination
8	$c \in A \Leftrightarrow c \in B$	4,7 $\Leftrightarrow$-introduction
9	$\forall x : X \bullet x \in A \Leftrightarrow x \in B$	8 $\forall$-introduction
10	$A = B$	9 Extension
11	$A \subseteq B \wedge B \subseteq A \Rightarrow A = B$	1-10 $\Rightarrow$-introduction
12	$\forall S, T : \mathbb{P}\, X \bullet$ $S \subseteq T \wedge T \subseteq S \Rightarrow S = T$	11 $\forall$-introduction

QED

Exercise 6.11 (Laws) *Prove the following laws of subset*

1. *Subset is transitive*

$$\forall S, T, U : \mathbb{P}\, X \bullet S \subseteq T \wedge T \subseteq U \Rightarrow S \subseteq U$$

2. *If S is a subset of T, then S is a member of the power set of T*

$$\forall S, T : \mathbb{P}\, X \bullet S \subseteq T \Leftrightarrow S \in \mathbb{P}\, T$$

3. *The empty set is a subset of every set*

$$\forall S : \mathbb{P}\, X \bullet \{\} \subseteq S$$

Exercise 6.12 (Laws) *Prove the following laws of proper subset*

1. *Proper subset is irreflexive*

$$\forall S : \mathbb{P}\, X \bullet \neg(S \subset S)$$

2. *Proper subset is asymmetric*

$$\forall S, T : \mathbb{P}\, X \bullet \neg(S \subset T \wedge T \subset S)$$

3. *Proper subset is transitive*

$$\forall S, T, U : \mathbb{P}\, X \bullet S \subset T \wedge T \subset U \Rightarrow S \subset U$$

4. *Every nonempty set has the empty set as a proper subset*

$$\forall S : \mathbb{P}\, X \bullet S \neq \{\} \Rightarrow \{\} \subset S$$

Exercise 6.13 *Define the superset and proper superset relations, and provide and prove correct a collection of appropriate laws.*

Exercise 6.14 *Prove the following assertions correct*

1. *The following are reflexive:* $\{\}, Id_A, =, \leq, \subseteq, \uparrow, A \times A$
2. *The following are irreflexive:* $\{\}, Tet, press, on, open, <$
3. *The following are symmetric:* $\{\}, Tet, =, A \times A$
4. *The following are antisymmetric:* $\{\}, \subseteq, \uparrow, =$
5. *The following are asymmetric:* $\{\}, \subset$
6. *The following are transitive:* $\{\}, \subset, \subseteq, <, =, A \times A$
7. *The following are equivalence relations:* $\{\}, =, A \times A$

Closures

In the last section we showed how new relations may be obtained by composing existing relations. In this section, we show how to create new relations that are supersets of existing ones, and have some general property. Let R be a homogeneous binary relation on A, that is

$$R : A \leftrightarrow A$$

then the *reflexive closure of* R is the smallest reflexive relation containing R. The reflexive closure of R is denoted R^r, and it enjoys the following properties

1. R^r is reflexive, that is, $Id_A \subseteq R^r$.

2. $R \subseteq R^r$.

3. R^r is the smallest such relation satisfying 1 and 2, that is, for any other reflexive relation S containing R, S also contains the reflexive closure R^r

$$R \subseteq S \Rightarrow R^r \subseteq S$$

We can define two similar relations, the *symmetric closure* of R, denoted R^s, and the *transitive closure* of R, denoted R^t.

Exercise 6.15 *Prove the following theorems correct:*

1. R *is* reflexive $\Leftrightarrow R = R^r$.

2. R *is* symmetric $\Leftrightarrow R = R^s$.

3. R *is* transitive $\Leftrightarrow R = R^t$.

The closures of a relation can be very useful in specification work; in particular, there is another kind of transitive closure that frequently arises: the *reflexive transitive closure* R^*

$$R^* \;\hat{=}\; \{x, y : S \mid x \mapsto y \in R^t \lor x \mapsto y \in Id_S \bullet x \mapsto y\}$$

Exercise 6.16 (Laws) *Prove the following laws of the closure operators:*

1. $\{\}^r = Id_A$
2. $Id_A^r = Id_A^s = Id_A^t = Id_A^* = Id_A$
3. $R^* \mathbin{⨾} R = R^t$
4. $R^t \mathbin{⨾} R = R^t = R \mathbin{⨾} R^*$
5. $(R^t)^t = R^t$
6. $(R^t)^{-1} = (R^{-1})^t$
7. $(R^*)^{-1} = (R^{-1})^*$
8. $R \mathbin{⨾} S = S \mathbin{⨾} R \Rightarrow (R \mathbin{⨾} S)^t = R^t \mathbin{⨾} S^t$

6.4 A Configuration Manager

At the heart of a Programming Support Environment is a mechanism to keep track of the various versions of program modules. Such a mechanism is sometimes called a configuration manager.

In this section we shall describe a configuration manager[4] that maintains information about modules written in a language such as Modula or Ada. In this language, modules have names drawn from the set *Name*. We need not dwell on the precise structure of the elements of *Name*: that would be an implementation matter. Instead, we shall leave *Name* as a parameter of the specification. Such a parameter is called a *given set*, and we regard it as a primitive type.

We shall not describe the actual code that goes into a module, as such a description isn't pertinent to our exposition. Rather, we shall introduce another given set *Body* of all (syntactically correct) module bodies.

---

[4] This specification is a simplification of one found in B. Sufrin and J. Woodcock, "Towards the Formal Specification of a Simple Programming Support Environment", *Software Engineering Journal*, July 1987.

```
t1  module ScreenHandler
    begin
       proc putchar
       ...
    end
t2  module Transput with ScreenHandler
    begin
       proc writestring
       ... putchar ...
    end
t3  module Application with Transput
    begin
       ... writestring ...
    end
```

Figure 6.2: Fragments of Program Modules

The interesting thing about a module is that it may import definitions (of program functions, procedures, types, etc.) from other modules. For example, consider the three fragments of program modules in Figure 6.2. The module *Application* needs to write a string of characters to the screen. Since this is such a common thing to do, there is a standard module which deals with "transput", and contains the definition of *writestring*. *Application* imports the definition from *Transput*. *Transput* in turn implements *writestring* in terms of another lower-level routine called *putchar*, which is imported from the first module, *ScreenHandler*.

So, at our level of abstraction the text of a module consists of a triple: the name of the module; a set of names of imported modules; and the body of the module. Common sense tells us that a module shouldn't import itself

$$Text : \mathbb{P}(Name \times (\mathbb{P}\, Name) \times Body)$$

$$Text = \{n : Name; s : \mathbb{P}\, Name; p : Body \mid n \notin s \bullet (n, s, p)\}$$

Our three concrete examples of elements of *Text* from Figure 6.2 are

$$
\begin{aligned}
t_1 = (&\mathit{ScreenHandler}, \{\},\\
&\quad \textbf{module } \mathit{ScreenHandler}\\
&\quad\quad \textbf{begin proc } \mathit{putchar} \cdots \textbf{end})\\
t_2 = (&\mathit{Transput}, \{\mathit{ScreenHandler}\},\\
&\quad \textbf{module } \mathit{Transput} \textbf{ with } \mathit{ScreenHandler}\\
&\quad\quad \textbf{begin proc } \mathit{writestring} \cdots \mathit{putchar} \cdots \textbf{end})\\
t_3 = (&\mathit{Application}, \{\mathit{Transput}\},\\
&\quad \textbf{module } \mathit{Application} \textbf{ with } \mathit{Transput}\\
&\quad\quad \textbf{begin} \cdots \mathit{writestring} \cdots \textbf{end})
\end{aligned}
$$

A Theory of Program Modules

We can describe the relationship between the text of a module and the names of the modules that it imports. It is as simple as it sounds: a text imports those modules named in its set of imports

$$\_\mathit{imports}\_ : \mathit{Text} \leftrightarrow \mathit{Name}$$

$$(n_1, s, p) \; \mathit{imports} \; n_2 \Leftrightarrow n_2 \in s$$

The following facts about the example in Figure 6.2 are not hard to verify

$$
\begin{aligned}
&t_1 \notin \mathit{dom\ imports}\\
&t_2 \; \mathit{imports} \; \mathit{ScreenHandler}\\
&t_3 \; \mathit{imports} \; \mathit{Transput}
\end{aligned}
$$

Another relationship that we can describe is built on *imports*, but it is a relation on *Texts*: we say that one text *needs* another if it imports the second's name

$$\_\mathit{needs}\_ : \mathit{Text} \leftrightarrow \mathit{Text}$$

$$(n_1, s_1, p_1) \; \mathit{needs} \; (n_2, s_2, p_2) \Leftrightarrow (n_1, s_1, p_1) \; \mathit{imports} \; n_2$$

Our common sense observation may be extended to this relation: no text needs itself, or, more formally, *needs* is irreflexive. We can prove that the irreflexivity of *needs* follows straight from the definition of what a module text is

Proof: Let $n_0 : Name$, $s_0 : \mathbb{P}\ Name$, *and* $b_0 : Body$ *be arbitrary*

1	$(n_0, s_0, p_0) \in Text$	assumption
2	$n_0 \notin s_0 \wedge$ $(n_0, s_0, p_0) \in$ $(Name \times (\mathbb{P}\ Name) \times Body)$	1Comprehension
3	$n_0 \notin s_0$	2 $\wedge$-elimination
4	$\neg(n_0 \in s_0)$	3 $\notin$-defn
5	$\neg((n_0, s_0, p_0)\ imports\ n_0)$	4 *imports*-defn
6	$\neg((n_0, s_0, p_0)\ needs\ (n_0, s_0, p_0))$	5 *needs*-defn
7	$(n_0, s_0, p_0) \in Text \Rightarrow$ $\neg((n_0, s_0, p_0)\ needs\ (n_0, s_0, p_0))$	1-6 $\Rightarrow$-introduction
8	$\forall t : (Name \times (\mathbb{P}\ Name) \times Body) \bullet$ $t \in Text \Rightarrow \neg(t\ needs\ t)$	7 $\forall$-introduction
9	$\forall t : Text \bullet \neg(t\ needs\ t)$	8 $\forall$-property
10	$\neg \exists t : Text \bullet t\ needs\ t$	$\exists$-theorem

QED

Figure 6.3: Proof that *needs* is irreflexive

Theorem 6.1 *The relation needs is irreflexive*

$$\neg \exists t : Text \bullet t\ needs\ t$$

The proof is displayed in Figure 6.3.

We can divide the set of module texts into what we shall call "compatible" modules. We can regard the name of a module and the set of module names that it imports as constituting the "signature" of a module. Two modules are compatible if they share the same signature

$$\_compat\_ : Text \leftrightarrow Text$$

$$(n_1, s_1, p_1)\ compat\ (n_2, s_2, p_2) \Leftrightarrow n_1 = n_2 \wedge s_1 = s_2$$

Exercise 6.17 *Prove that compat is an equivalence relation on Text.*

The State

We shall continue the specification of the configuration manager by describing its *state*. The state consists of those data structures that are maintained by the system; the operations in the system manipulate these data structures, and therefore change the state. We are going to use mathematical data types to build a theory of the state which will have as models suitable realisations in a programming language. We shall describe how the state may be initialised, and how values of the state before and after operations are related.

The state contains two components: a *store* of *Texts*, and a relation which describes when one stored text is a version which is a *successor* for another stored text

$$store : \mathbb{P}\ Text$$
$$\_suc\_ : Text \leftrightarrow Text$$

We can constrain the successor relation in the following ways. A module must never be the successor of itself, for that would introduce an unbreakable "loop" if we tried to find the end of the chain. Thus the *transitive closure* of *suc* is irreflexive

$$suc^t \in Irreflexive[Text]$$

where $Irreflexive[X]$ is the set of all irreflexive relations on X

$$Irreflexive[X] \;\hat{=}\; \{R : X \leftrightarrow X \mid (\neg \exists x : X \bullet x\ R\ x)\}$$

If a text t_1 is the successor of another text t_2, then they must be compatible, that is *suc* is merely a subset of *compat*

$$suc \subseteq compat$$

All the nominated modules that have, or are themselves, successors must reside in the store

$$suc \subseteq store \times store$$

store and *suc* constitute a proper state of the configuration manager whenever they satisfy these three constraints

$$suc^t \in Irreflexive[Text]$$
$$suc \subseteq compat$$
$$suc \subseteq store \times store$$

An Introduction to Schemas

The mathematical language that we have introduced is sufficiently powerful to describe many aspects of software systems. However, its application to large-scale specification work soon results in unwieldy descriptions that are difficult to follow. It isn't the mathematical language that is at fault, but rather our human need to comprehend just a small amount of information at a time. Therefore, we must present mathematical descriptions in a sympathetic fashion, explaining small parts in the simplest possible context, and then showing how to fit the pieces together to make the whole.

One of the most basic things that we can do to help the reader—or indeed the writer—of a specification is to identify and name commonly used concepts and factor them out from the rest of the description of a system. In this way, we can encapsulate an important concept and give it a name, thus increasing our vocabulary—and our mental power!

In specifications, we see a pattern occurring over and over again: a piece of mathematical *structure* which describes some constrained variables. We call this introduction of variables under some constraint a *schema*.

Example 6.13 *The following set comprehension term and quantified predicate each contain the same pattern of constrained variables*

$$\{m, n : \mathbb{N} \mid n = 2 \times m \bullet m \mapsto n\}$$
$$\forall x, y : \mathbb{N} \mid x \neq y \bullet x > y \vee y > x$$

The pattern is

$$declaration \mid predicate$$

So, for these two examples we have the patterns

$$m, n : \mathbb{N} \mid n = 2 \times m$$
$$x, y : \mathbb{N} \mid x \neq y$$

Returning to our configuration management example, the three predicates that characterise the state of the configuration manager together form *the state invariant.* A set of texts and a relation on texts do not constitute a possible state of the configuration manager unless they satisfy the state invariant.

The state invariant, together with the declarations of *store* and *suc*, form a schema which we shall call *ConfigMan*. This package of definitions is a mathematical structure that describes all legal values of the state of the configuration manager

ConfigMan

$store : \mathbb{P}\ Text;$
$\_suc\_ : Text \leftrightarrow Text$

$suc^t \in Irreflexive[Text] \wedge$
$suc \subseteq compat \wedge$
$suc \subseteq store \times store$

A schema consists of two parts

- a *declaration* of some variables; and
- a *predicate* constraining their values.

In order to reduce unnecessary formal clutter, we often put declarations on separate lines and omit the semicolon. Similarly, we often put conjuncts on separate lines and omit the conjunction symbol

ConfigMan

$store : \mathbb{P}\ Text$
$\_suc\_ : Text \leftrightarrow Text$

$suc^t \in Irreflexive[Text]$
$suc \subseteq compat$
$suc \subseteq store \times store$

This schema is *equivalent* to that previously presented. Remember, each line of a predicate forms a conjunct, so

$a \Rightarrow b$
c
$d \vee e$

means

$$(a \Rightarrow b) \wedge c \wedge (d \vee e)$$

It doesn't matter in which order we write the declarations in a schema's declaration part. We would be very surprised indeed if the following schema defined something different from the one that we referred to as *ConfigMan*

$$
\begin{array}{|l}
\hline
\textit{ConfigMan} \\
\_suc\_ : Text \leftrightarrow Text \\
store : \mathbb{P}\ Text \\
\hline
suc \subseteq store \times store \\
suc \subseteq compat \\
suc^t \in Irreflexive[Text] \\
\hline
\end{array}
$$

Naming a schema introduces the name as a syntactic abbreviation for the schema. Schemas are named by embedding the name in the top line of the schema's box

$$
\begin{array}{|l}
\hline
\textit{Name} \\
declarationpart \\
\hline
predicate \\
\hline
\end{array}
$$

We can always replace the name of a schema by the mathematical structure it refers to. When we describe changes to the state of the configuration manager we shall describe a relation between two states: the state *before* the operation and the state *after* the operation. A convention that we shall adopt is that we shall decorate the names of "after" variables with a dash to distinguish them from "before" variables.

Example 6.14 *Having described the state of the configuration manager, we must say what the* initial state *of the system is. The initial state of the configuration manager is an empty store and an empty successor relation. We can regard the initialisation of a system as a peculiar kind of operation that creates a state out of nothing; there is no before state, simply an after state, with its variables decorated. We* *call the initialisation operation InitConfigMan, and it characterises what it is to be an initial state of the configuration management system*

$$\begin{array}{|l}
\hline
\textit{InitConfigMan} \\
store' : \mathbb{P}\ Text \\
suc' : Text \leftrightarrow Text \\
\hline
store' = \{\} \\
suc' = \{\} \\
\hline
\end{array}$$

There is only one state of the configuration manager that satisfies this description; it wouldn't matter if there were more, but there must be at least one, or we could never implement the system. There is an obligation upon us to prove that an initial state exists. Wherever possible, we like to have concise descriptions, so that the reader's attention is focused on important details, rather than irrelevancies. If we decorate the name of a schema, we understand it to mean the same mathematical structure, but with the component names so decorated. The state after an operation is described by ConfigMan'. Expanding this we get

$$\begin{array}{|l}
\hline
\textit{ConfigMan}' \\
suc' : Text \leftrightarrow Text \\
store' : \mathbb{P}\ Text \\
\hline
suc' \subseteq store' \times store' \\
suc' \subseteq compat \\
suc'^{t} \in Irreflexive[Text] \\
\hline
\end{array}$$

This allows us a concise expression of our proof obligation in the Initialisation Theorem.

Theorem 6.2 (Initialisation) *An initial state for the configuration manager exists*

$$\exists\ ConfigMan' \bullet InitConfigMan$$

Proof: This follows from the various laws concerning the empty set and the operators that we have introduced so far, and is contained in Figure 6.4 on page 116.

Exercise 6.18 *In our example we haven't fully explained what a successor to a text t is: the successor to t is always preferred to t. This doesn't make a lot of sense unless each text has at most* one *successor. Rewrite the description of the configuration manager to reflect this requirement: you should describe the revised state, the revised initial state, and you should prove the initialisation theorem.*

Exercise 6.19 *A* variant *of a text is rather like a successor, in that it must have the same signature, but the body is different. However, a variant is not* always *preferred, it differs for some particular reason. For example, we might have two texts that are variants of each other, one is designed for space efficiency, the other for speed efficiency. Develop a theory of variants and add a suitable relation to the state of the configuration manager.*

Exercise 6.20 *Write down the names of the operations that you would like to find in the interface to the configuration manager.*

Exercise 6.21 *Describe the state of a system of your choice using the notation that we have introduced so far in the book. Describe the initial state of your system and prove its initialisation theorem.*

6.5 Summary

The notation introduced in this chapter has increased our ability to capture specifications of software systems to the extent that we concluded with the specification of a modest, but nonetheless realistic, system. We have described what a *relation* is, and started to describe its vital importance in the modelling of state-based systems. We have described some useful properties of relations that allow us to categorise them, and some useful combinators that allow us to construct new relations from old ones. Finally, in our case study, we have started to introduce a way of presenting the mathematics that we use to specify software: the *schema*.

1	$\{\} \in \mathbb{P}\ Text$	$\in$-property
2	$\{\} \in Text \leftrightarrow Text$	$\in$-property
3	$\{\} \in Irreflexive[Text]$	*Irreflexive*-property
4	$\{\}^t = \{\}$	R^t-property
5	$\{\}^t \in Irreflexive[Text]$	3,4 substitution
6	$\{\} \subseteq compat$	$\subseteq$-property
7	$\{\} \subseteq \{\}$	$\subseteq$-property
8	$\{\} \times \{\} = \{\}$	$\times$-property
9	$\{\} \subseteq \{\} \times \{\}$	7,8-substitution
10	$\{\} \in \mathbb{P}\ Text \land$ $\{\} \in Text \leftrightarrow Text \land$ $\{\}^t \in Irreflexive[Text] \land$ $\{\} \subseteq compat \land$ $\{\} \subseteq \{\} \times \{\}$	1,2,5,6,9 $\land$-introduction
11	$\exists\, store' : \mathbb{P}\ Text;$ $suc' : Text \leftrightarrow Text \bullet$ $suc'^t \in Irreflexive[Text] \land$ $suc' \subseteq compat \land$ $suc' \subseteq store' \times store' \land$ $store' = \{\} \land$ $suc' = \{\}$	10 $\exists$-introduction
12	$\exists\, ConfigMan' \bullet$ $store' = \{\} \land$ $suc' = \{\}$	11-substitution
13	$\exists\, ConfigMan' \bullet InitConfigMan$	12-substitution
QED		

Figure 6.4: Proof of the Initialisation Theorem.

Chapter 7

Functions

This chapter introduces the notion of a mathematical function. Many examples of functions are given as we build up a useful collection of operators on the constructs that we have introduced in previous chapters, namely sets and relations. To illustrate the usefulness of all this notation, a case study is included which develops a small system which keeps track of how resources are being used by a group of users.

7.1 An Introduction to Functions

An idea central to both mathematics and computation is that of *function*. A function is something that gives a deterministic answer to some question. It is deterministic in the sense that it always replies with the same answer to a particular question. If asked "What is the square root of 4?", we would feel justified in giving either of the answers "+2" or "-2". In fact, we could be capricious and vary the choice between the answers in order on each occasion so as to cause maximum inconvenience to the questioner[1]. We certainly couldn't be relied upon. The point is, "What is the square root?" is not a function, but "What is the positive square root?" and "What is the negative square root?" both are functions.

We define a *partial function* f from A to B to be something that

[1] This is a particular kind of nondeterminism called *demonic nondeterminism*.

maps an element from A onto at most one element of B. We denote the set of all such partial functions from a set A to a set B

$$A \rightarrowtail\!\!\!\!\!\!+\; B$$

These functions are partial in the sense that there is no necessity for all the members of A to be mapped to members of B, the function might be defined on only a subset of A.

Example 7.1 *Let psqrt be the partial function mapping numbers into their exact, positive square roots. Then,*

$$psqrt : \mathbb{N} \nrightarrow \mathbb{N}$$

This is a partial function because, for example, 7 *is not in the domain of psqrt.*

We denote the application of a function to an element taken from its domain by juxtaposition, thus[2]

$$psqrt\ 9 = 3$$

We can describe the function *psqrt* by saying to what it maps its elements

$$\forall m, n : \mathbb{N} \bullet (f\ m = n) \Leftrightarrow (m = n \times n)$$

We have defined *psqrt* by saying what kind of function it is ($\mathbb{N} \nrightarrow \mathbb{N}$) and what it does in order to achieve its effect; we can put these together as

$$\begin{array}{|l} psqrt : \mathbb{N} \nrightarrow \mathbb{N} \\ \hline \forall m, n : \mathbb{N} \bullet psqrt\ m = n \Leftrightarrow m = n \times n \end{array}$$

We can view a function as a specialised kind of relation—any relation that is deterministic (that is, functional)

$$A \nrightarrow B \mathrel{\widehat=} \{R : A \leftrightarrow B \mid \forall x : A;\ y, z : B \bullet x\ R\ y \wedge x\ R\ z \Rightarrow y = z\}$$

[2]Some authors, unaware of the acute international shortage of parentheses, squander them and write *psqrt*(9), which of course means the same thing.

Using this view, we can explain function application as follows: for

$$f \in A \nrightarrow B$$

if x is in the domain of f, then $f\ x$ stands for that element of B to which f relates x. That is,

$$x \in dom\ f \Rightarrow (x \mapsto f\ x) \in f$$

where *dom* f is the domain[3] of f. This small piece of insight allows us to treat functions just as special cases of relations.

As an example, let

$$\begin{aligned} U &\triangleq \{u_1, u_2, \ldots, u_{50}\} \\ B &\triangleq \{1, 2, \ldots, 100\} \end{aligned}$$

then we can define a function from B to U

$$dir : B \nrightarrow U$$

$$dir = \{1 \mapsto u_2, 3 \mapsto u_4, 4 \mapsto u_4, 5 \mapsto u_4, 10 \mapsto u_7\}$$

That *dir* is a relation between B and U is quite clear; we have, however, an obligation to show that it is also functional. Happily, this is easy, since evidently each element of its domain is mapped onto precisely one element of its range. It is a partial function, since we can find an element of B, for example 99, which is not mapped onto anything in U. Whenever we define something and claim that it is a function, we should carry out such an argument.

Figure 7.1 shows *dir* represented as a picture. The domain of *dir* is on the left of the picture, and the range on the right. The maplets are represented by arrows from domain elements to range elements. As the domain and range are both fairly large, we have only included those elements in the picture that are actually involved in the relationship that *dir* represents. In such a picture, the functionality of the relation is captured by noting that there are no *diverging* arrows.

[3] We had an informal definition of *dom* f in Chapter 6; we shall shortly come to a formal definition of *dom* f.

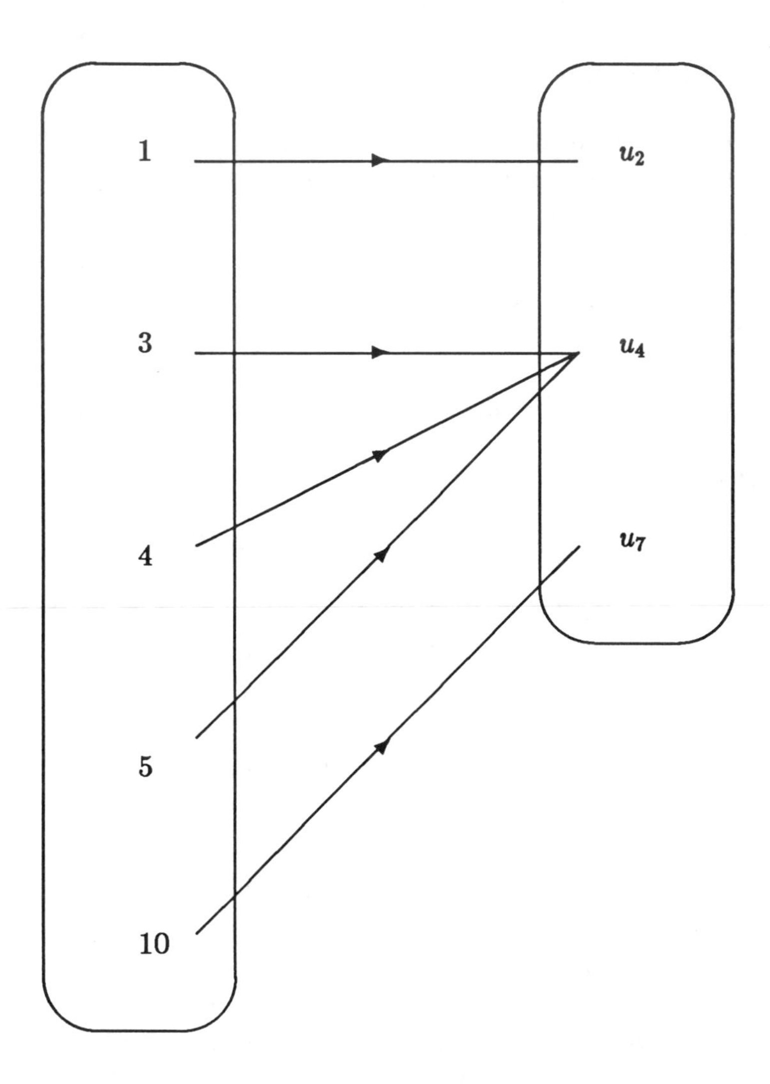

Figure 7.1: The function *dir*.

Since 3 is in the domain of *dir*, we can safely apply *dir* to it, thus

$$dir\ 3 = u_4$$

If we are foolish enough to apply *dir* to an element not in its domain, then what we have written is meaningless.

Sometimes it is important that the domain of a function

$$f : A \nrightarrow B$$

is the *whole* of A. We call such a function a *total function* from A to B, and write

$$f : A \rightarrow B$$

Again, taking $dom\ f$ to be the domain of f, we have

$$A \rightarrow B \ \widehat{=}\ \{f : A \nrightarrow B \mid dom\ f = A\}$$

Clearly,

$$A \rightarrow B \subseteq A \nrightarrow B$$

Let's define another object

$$\begin{array}{|l} bigdir : B \rightarrow U \\ \hline bigdir = \\ \quad \{b : B;\ u : U \mid \\ \qquad (b \in even \wedge u = u_1) \vee (b \in odd \wedge u = u_2) \bullet b \mapsto u\} \end{array}$$

Exercise 7.1 *Write out the maplets in the function bigdir with b between* 1 *and* 10.

Now the fact that *bigdir* is functional follows from the observation that no number can be simultaneously even and odd

$$\neg \exists n : \mathbb{N} \bullet n \in even \wedge n \in odd$$

That *bigdir* is total follows from the fact that every number in B is either even or odd

$$\forall n : B \bullet n \in even \vee n \in odd$$

When we define a function we can use the techniques that we used to define relations: we can define it by enumerating the constituent maplets (as we did with *dir*), or by defining the function as a set in comprehension (as we did with *bigdir*). Either way, we must remember our obligation to prove that the relation so defined is actually *functional.* As we shall shortly see, there are many ways of combining existing functions and sets to form new functions. The proof obligation usually remains.

Example 7.2 *Here are two functions that return the minimum and maximum of a set of numbers*

$$min, max : \mathbb{P}\,\mathbb{N} \mathrel{+\!\!\!\!\rightarrow\!\!\!\!\!\rightarrow} \mathbb{N}$$

$$\begin{aligned} min &= \{S : \mathbb{P}\,\mathbb{N};\ m : \mathbb{N} \mid \\ &\qquad S \neq \{\} \land m \in S \land (\forall i : S \bullet m \leq i) \bullet S \mapsto m\} \\ max &= \{S : \mathbb{P}\,\mathbb{N};\ m : \mathbb{N} \mid \\ &\qquad S \neq \{\} \land m \in S \land (\forall i : S \bullet m \geq i) \bullet S \mapsto m\} \end{aligned}$$

Exercise 7.2 *Prove that min and max are functions.*

There is another way of presenting functions that we shall use later in the book which the reader will often find in the literature: it is the *lambda abstraction.* This notation often looks wonderfully mystical, but alas we have no need for such mysticism, and we shall treat it merely as another way of recording a function. The structure of a lambda expression is

$$\lambda\langle arguments\rangle \mid \langle constraint\rangle \bullet \langle result\rangle$$

The *constraint* on the arguments is optional. The following are examples of lambda expressions.

Example 7.3 $\lambda\, n : \mathbb{N} \bullet n + n$ *is a total function of* $\mathbb{N} \rightarrow \mathbb{N}$ *which returns as its result twice the value of its argument. It is equivalent to the expression*

$$\{n : \mathbb{N} \bullet n \mapsto (n + n)\}$$

We can give the lambda expression a name

$$double \mathrel{\widehat{=}} \lambda\, n : \mathbb{N} \bullet n + n$$

and apply it to an actual argument

$$double\ 4 = (\lambda\, n : \mathbb{N} \bullet n + n)\ 4 = 8$$

Example 7.4 *The role of the constraint is to limit the domain of application of the function.*

$$\lambda\, n : \mathbb{N} \mid n \neq 0 \bullet n - 1$$

describes a partial *function*

$$\mathbb{N} \nrightarrow \mathbb{N}$$

which returns as its result one less than its argument. It is partial because the value 0 is specifically excluded as an argument by the constraint. It is equivalent to the set comprehension expression

$$\{n : \mathbb{N} \mid n \neq 0 \bullet n \mapsto (n - 1)\}$$

Example 7.5 *The argument to a function may be a pair of values— or in fact any complicated structure*

$$\lambda\, m, n : \mathbb{N} \bullet (double\ n, double\ m)$$

is a function which takes a pair of arguments

$$\mathbb{N} \times \mathbb{N} \rightarrow \mathbb{N} \times \mathbb{N}$$

and it describes the function

$$\{m, n : \mathbb{N} \bullet (m, n) \mapsto (double\ n, double\ m)\}$$

which doubles each of its two arguments and transposes them. For example

$$(\lambda\, m, n : \mathbb{N} \bullet (double\ n, double\ m))\ (3, 4) = (8, 6)$$

Example 7.6 *The result of applying a lambda expression can be any kind of object, as we see in this example, where a function is returned*

$$pair\ \widehat{=}\ \lambda f, g : \mathbb{N} \rightarrow \mathbb{N} \bullet (\lambda\, x : \mathbb{N} \bullet (f\ x, g\ x))$$

This function takes two functions as arguments and returns a third [4]

$$((\mathbb{N} \to \mathbb{N}) \times (\mathbb{N} \to \mathbb{N})) \to (\mathbb{N} \to (\mathbb{N} \times \mathbb{N}))$$

The result of applying pair to two functions as arguments can be seen in the expression

$$\begin{aligned} & pair(double, double) \\ & \quad = (\lambda f, g : \mathbb{N} \to \mathbb{N} \bullet (\lambda x : \mathbb{N} \bullet (f\ x, g\ x)))(double, double) \\ & \quad = \lambda x : \mathbb{N} \bullet (double\ x, double\ x) \end{aligned}$$

We can take this function and apply it to an argument in the normal way

$$\begin{aligned} & (pair(double, double))\ 15 \\ & \quad = (\lambda x : \mathbb{N} \bullet (double\ x, double\ x))\ 15 \\ & \quad = (double\ 15, double\ 15) \\ & \quad = (30, 30) \end{aligned}$$

7.2 A Mathematical Rattle-Bag

Functions turn out to be extremely useful objects. In this section we consider functions that manipulate the basic entities that were introduced in the previous sections: sets, relations, and other functions.

Operations on Sets

When we learn about something new, such as sets, we are naturally curious about what happens when we combine these new things in various ways—the *algebra* for the objects. The first three such operations on sets that we consider are rather like the simple operations on numbers of *addition*, *multiplication*, and *subtraction*.

[4] In mathematics it is generally agreed that the function arrow is *right associative*: $A \to B \to C$ means $A \to (B \to C)$. Function application on the other hand is *left associative*: $f\ x\ y$ means $(f\ x)\ y$. If there are to be conventions such as these, then it is wise that the convention for the associativity of function arrows is the dual of that for function applications.

Set Union

The *union* of two sets S and T is the set of all elements which belong to S or to T or to both. We denote S *united with*[5] T by

$$S \cup T$$

As examples, consider the following

$$\{1\} \cup \{2,3\} = \{1,2,3\}$$
$$\{\} \cup \{1,2,3\} = \{1,2,3\}$$
$$\{1,2,3\} \cup \{\} = \{1,2,3\}$$
$$\{1,2\} \cup \{2,3\} = \{1,2,3\}$$
$$\{1,2,3\} \cup \{1,2,3\} = \{1,2,3\}$$

We can formalise this intuitive notion of union as follows. If we take any set X, then

$$[X]$$
$$\_\cup\_ : \mathbb{P}\, X \times \mathbb{P}\, X \rightarrow \mathbb{P}\, X$$
$$\forall S, T : \mathbb{P}\, X \bullet$$
$$S \cup T = \{x : X \mid x \in S \lor x \in T\}$$

For example, we can rewrite the predicate

$$\forall n : B \bullet n \in even \lor n \in odd$$

as an expression using sets

$$B \subseteq even \cup odd$$

Notice the connection between set union and disjunction, and also that we can unite two sets only if they are of the same type, nothing else makes sense.

Exercise 7.3 *Prove that the following laws hold for set union*

$S \cup S = S$	*idempotence*
$S \cup T = T \cup S$	*commutativity*
$S \cup (T \cup U) = (S \cup T) \cup U$	*associativity*
$S \cup \{\} = S$	*identity element*
$S \subseteq (S \cup T)$	

[5] There is no verb "to *union*".

Set Intersection

The intersection of two sets S and T is the set of all elements which are common to both S and T. We denote S *intersected with* T by

$$S \cap T$$

As examples, consider

$$\{\} \cap \{1,2,3\} = \{\}$$
$$\{1,2,3\} \cap \{\} = \{\}$$
$$\{1\} \cap \{2,3\} = \{\}$$
$$\{1,2\} \cap \{2,3\} = \{2\}$$
$$\{1,2,3\} \cap \{1,2,3\} = \{1,2,3\}$$

Just like set union, set intersection is also a function from a pair of sets to a third set For any set X

$$[X]$$
$$\_\cap\_ : \mathbb{P}\, X \times \mathbb{P}\, X \to \mathbb{P}\, X$$
$$\forall S, T : \mathbb{P}\, X \bullet$$
$$\quad S \cap T = \{x : X \mid x \in S \wedge x \in T\}$$

Notice the connection between set intersection and conjunction, and that we can intersect two sets only if they are of the same type. We can also express the predicate

$$\neg \exists n : \mathbb{N} \bullet n \in \mathit{even} \wedge n \in \mathit{odd}$$

in terms of sets

$$\mathit{even} \cap \mathit{odd} = \{\}$$

This says that the intersection of *even* and *odd* is the empty set—they don't overlap. The mathematician's phrase for this is that they are *disjoint*.

Exercise 7.4 *Prove that the following laws hold for set intersection*

$S \cap S = S$	*idempotence*
$S \cap T = T \cap S$	*commutativity*
$S \cap (T \cap U) = (S \cap T) \cap U$	*associativity*
$S \cap \{\} = \{\}$	*zero element*
$(S \cap T) \subseteq S$	

Exercise 7.5 *We also find that union and intersection are* mutually distributive, *that is*

$$S \cup (T \cap U) = (S \cup T) \cap (S \cup U)$$
$$S \cap (T \cup U) = (S \cap T) \cup (S \cap U)$$

which we would expect from observing the connections with disjunction and conjunction, respectively. Prove these identities.

Set Difference

The *difference* of two sets S and T is the set of all elements which belong to S, but do not belong to T. We denote the difference of S and T by

$$S \setminus T$$

As examples, consider

$$\{1,2,3\} \setminus \{1,2\} = \{3\}$$
$$\{\} \setminus \{1,2,3\} = \{\}$$
$$\{1,2,3\} \setminus \{\} = \{1,2,3\}$$
$$\{1\} \setminus \{2,3\} = \{1\}$$
$$\{1,2\} \setminus \{2,3\} = \{1\}$$
$$\{1,2,3\} \setminus \{1,2,3\} = \{\}$$

For any set X

$$\begin{array}{l} [X] \\ \hline \_\setminus\_ : \mathbb{P}\,X \times \mathbb{P}\,X \rightarrow \mathbb{P}\,X \\ \hline \forall S, T : \mathbb{P}\,X \bullet \\ \qquad S \setminus T = \{x : X \mid x \in S \land x \notin T\} \end{array}$$

Exercise 7.6 *Prove that the following laws hold for set difference*

$$S \setminus S = \{\}$$
$$S \setminus \{\} = S$$
$$\{\} \setminus S = \{\}$$
$$(S \setminus T) \setminus U = S \setminus (T \cup U)$$
$$(S \setminus T) \subseteq S$$

Exercise 7.7 *Set difference also distributes backwards through both union and intersection*[6]

$$(S \cup T) \setminus U = (S \setminus U) \cup (T \setminus U)$$
$$(S \cap T) \setminus U = (S \setminus U) \cap (T \setminus U)$$

Exercise 7.8 *Prove the following theorems of set algebra:*

$$S \subseteq T \Rightarrow S \cap T = S$$
$$S \subseteq T \Rightarrow S \cup T = T$$
$$S \subseteq T \Rightarrow S \cup (T \setminus S) = T$$

Distributed Union

We have defined union and intersection as *binary* operators: the two arguments they take and the result that they produce are of the same type. We can be more adventurous and define generalised operators that take an arbitrary number of arguments and produce a result, all of the same type. We can collect the arguments into a set and thereby avoid being too specific about how many arguments there are. The fact that set union is both *commutative* and *associative* means that the order in which the arguments are united is unimportant. The *distributed union* of a set of sets SS is the set of all elements which are members of some member of SS. Examples are

$$\bigcup\{\{1,2,3\},\{2,3,4\},\{5,6\}\} = \{1,2,3,4,5,6\}$$
$$\bigcup\{\} = \{\}$$
$$\bigcup\{\{\}\} = \{\}$$
$$\bigcup\{\{1\},\{2\},\{3\}\} = \{1,2,3\}$$
$$\bigcup\{\{\},\{1,2,3\}\} = \{1,2,3\}$$

Formally, distributed union is defined for any set X as

[6] This is sometimes also called *right* distribution, because the set difference operator is being distributed through the union operator, and it appears on the right of the expression. It is also called *left* distribution because the distribution is moving towards the left. It seems worthwhile avoiding the evident confusion and using the terms *forwards* and *backwards* distribution. We hope that this is clearer.

$$
\begin{array}{l}
[X] \\
\hline
\bigcup : \mathbb{P}(\mathbb{P}\,X) \to \mathbb{P}\,X \\
\hline
\forall SS : \mathbb{P}(\mathbb{P}\,X) \bullet \\
\quad \bigcup SS = \{x : X \mid \exists S : SS \bullet x \in S\}
\end{array}
$$

Exercise 7.9 *Prove that distributed union enjoys the following laws*

$$
\begin{array}{l}
\bigcup\{\} = \{\} \\
\bigcup(SS \cup TT) = (\bigcup SS) \cup (\bigcup TT) \\
\bigcup\{S, T\} = S \cup T \\
(\bigcup SS) \setminus T = \bigcup\{S : SS \mid S \setminus T\} \\
SS \subseteq TT \Rightarrow \bigcup SS \subseteq \bigcup TT
\end{array}
$$

Distributed Intersection

The ***distributed intersection*** of a set of sets SS is the set of all elements which are members of all members of SS. Examples are

$$
\begin{array}{l}
\bigcap\{\{1\}, \{2\}, \{3\}\} = \{\} \\
\bigcap\{\{\}, \{1, 2, 3\}\} = \{\} \\
\bigcap\{\{1\}, \{1, 2\}, \{1, 2, 3\}\} = \{1\}
\end{array}
$$

Distributed intersection is defined for any set X as

$$
\begin{array}{l}
[X] \\
\hline
\bigcap : \mathbb{P}(\mathbb{P}\,X) \to \mathbb{P}\,X \\
\hline
\forall SS : \mathbb{P}(\mathbb{P}\,X) \bullet \\
\quad \bigcap SS = \{x : X \mid \forall S : SS \bullet x \in S\}
\end{array}
$$

Exercise 7.10 *Prove that distributed intersection enjoys the following laws, where X is any set*

$$
\begin{array}{l}
\bigcap\{\} = X \\
\bigcap(SS \cap TT) = (\bigcap SS) \cap (\bigcap TT) \\
\bigcap\{S, T\} - S \cap T \\
S \setminus (\bigcap TT) = \bigcap\{T : TT \mid S \setminus T\} \\
SS \subseteq TT \Rightarrow \bigcap SS \subseteq \bigcap TT
\end{array}
$$

The first of the laws in Exercise 7.10 deserves some examination. The distributed intersection over the empty set gives rise to a universal quantification over the empty set, which is vacuously true; thus every element of X satisfies the predicate in the definition, by default as it were. We are perhaps sometimes surprised at the consequences of our definitions.

Operations on Relations

Domain of a Relation

The *domain* of a relation R is the set of all elements which are related to something by R. Recall that we had

$$dir = \{1 \mapsto u_2, 3 \mapsto u_4, 4 \mapsto u_4, 5 \mapsto u_4, 10 \mapsto u_7\}$$

The domain of *dir* is

$$dom\ dir = \{1, 3, 4, 5, 10\}$$

If R relates things in X to things in Y, then $dom\ R$ is a subset of X

$$[X, Y]$$
$$dom : (X \leftrightarrow Y) \rightarrow \mathbb{P}\, X$$
$$\forall R : X \leftrightarrow Y \bullet$$
$$\quad dom\ R = \{x : X \mid \exists y : Y \bullet x\ R\ y\}$$

dom is a function from the set of relations from X to Y to $(\mathbb{P}\ X)$: we apply it to a function and it returns a set. This is an example of what is called a *higher order function*, since it takes a function as an argument. Some people make a lot of fuss about this, but since we regard functions merely as sets with knobs on, *dom* is just taking a set as its argument.

Exercise 7.11 *Prove that dom satisfies the following three laws*

$$dom\ \{\} = \{\}$$
$$dom\ \{x \mapsto y\} = \{x\}$$
$$dom\ (S \cup T) = (dom\ S) \cup (dom\ T)$$

Range of a Relation

The *range* of a relation R is the set of all elements which have something related to them by R. The range of *dir* is

$$ran\ dir = \{u_2, u_4, u_7\}$$

If R relates things in X to things in Y, then *ran* R is a subset of Y

$$[X, Y]$$
$$ran : (X \leftrightarrow Y) \rightarrow \mathbb{P}\ Y$$
$$\forall R : X \leftrightarrow Y \bullet$$
$$ran\ R = \{y : Y \mid \exists x : X \bullet x\ R\ y\}$$

Exercise 7.12 *Prove that ran satisfies the following three laws*

$$ran\ \{\} = \{\}$$
$$ran\ \{x \mapsto y\} = \{y\}$$
$$ran\ (S \cup T) = (ran\ S) \cup (ran\ T)$$

Domain Restriction

The following four functions are intended to allow us to manipulate relations. The expression

$$S \lhd R$$

denotes the relation formed by taking the relation R and restricting its domain to be wholly within the set S. Referring back to *dir* once more, if we restrict the function's domain to the set

$$\{3, 4, 5, 6\}$$

then we eliminate those maplets whose source is not 3, 4, 5, or 6, namely

$$1 \mapsto u_2 \text{ and } 10 \mapsto u_7$$

Thus

$$\begin{aligned}&\{3,4,5,6\} \lhd dir \\ &\quad = \{3,4,5,6\} \lhd \{1 \mapsto u_2, 3 \mapsto u_4, 4 \mapsto u_4, 5 \mapsto u_4, 10 \mapsto u_7\} \\ &\quad = \{3 \mapsto u_4, 4 \mapsto u_4, 5 \mapsto u_4\}\end{aligned}$$

The formal definition describes precisely the meaning of the operator

$$\begin{array}{|l} [A, B] \\ \_ \lhd \_ : (\mathbb{P}\, A) \times (A \leftrightarrow B) \rightarrow (A \leftrightarrow B) \\ \hline \forall S : \mathbb{P}\, A; R : A \leftrightarrow B \bullet \\ \qquad S \lhd R = \{a : A; b : B \mid a \in S \wedge a\ R\ b \bullet a \mapsto b\} \end{array}$$

The relation formed by domain restriction respects the original relation. In order to survive two successive restrictions, an element must be able to survive both of them at the same time. This gives rise to the following laws

$$\begin{aligned}&(S \lhd R) \subseteq R \\ &T \lhd (S \lhd R) = (T \cap S) \lhd R\end{aligned}$$

The domain of a function f whose domain is restricted by a set S must be contained in both S and the domain of f

$$dom\ (S \lhd f) = S \cap (dom\ f)$$

Exercise 7.13 *Prove the laws of domain restriction*

Range Restriction

Range restriction is very similar to domain restriction, but the syntax is swapped around to suggest the difference.

$$\begin{aligned}&dir \rhd \{u_2, u_7\} \\ &\quad = \{1 \mapsto u_2, 3 \mapsto u_4, 4 \mapsto u_4, 5 \mapsto u_4, 10 \mapsto u_7\} \rhd \{u_2, u_7\} \\ &\quad = \{1 \mapsto u_2, 10 \mapsto u_7\}\end{aligned}$$

Not surprisingly, the formal definition is symmetrical to that for domain restriction

$$
\begin{array}{|l}
[A, B] \\
\_ \rhd \_ : (A \leftrightarrow B) \times (\mathbb{P}\, B) \rightarrow (A \leftrightarrow B) \\
\hline
\forall R : A \leftrightarrow B; S : \mathbb{P}\, B \bullet \\
\quad R \rhd S = \{a : A; b : B \mid a\ R\ b \land b \in S \bullet a \mapsto b\}
\end{array}
$$

We have two familiar laws

$$(R \rhd S) \subseteq R$$
$$(R \rhd S) \rhd T = R \rhd (S \cap T)$$

The range of a function f whose range is restricted by a set S must be contained in both S and the range of f

$$\mathit{ran}\ (f \rhd S) = (\mathit{ran}\ f) \cap S$$

When we restrict both the domain and the range of a relation, it doesn't matter in which order it is done; this is rather like an associativity law

$$(S \lhd R) \rhd T = S \lhd (R \rhd T)$$

In future we shall omit the parentheses in such expressions, since they only clutter.

Exercise 7.14 *Prove the laws of range restriction.*

Domain Subtraction

Sometimes we wish to express a restriction in a more negative way, by saying that we want a relation that doesn't have any elements from a particular set in its domain. Domain subtraction is such an operator. The expression

$$S \ntriangleleft R$$

is the relation which is like R, but has no element from S in its domain.

$$
\begin{aligned}
&\{1, 10\} \ntriangleleft \mathit{dir} \\
&\quad = \{1, 10\} \ntriangleleft \{1 \mapsto u_2, 3 \mapsto u_4, 4 \mapsto u_4, 5 \mapsto u_4, 10 \mapsto u_7\} \\
&\quad = \{3 \mapsto u_4, 4 \mapsto u_4, 5 \mapsto u_4\} \\
&\quad = \{3, 4, 5, 6\} \lhd \mathit{dir}
\end{aligned}
$$

We define it to be

$$
\begin{array}{|l}
\hline
[A, B] \\
\_ \mathbin{⩤} \_ : (\mathbb{P}\, A) \times (A \leftrightarrow B) \rightarrow (A \leftrightarrow B) \\
\hline
\forall S : \mathbb{P}\, A; R : A \leftrightarrow B \bullet \\
\quad S \mathbin{⩤} R = \{a : A; b : B \mid a \notin S \land a\ R\ b \bullet a \mapsto b\} \\
\hline
\end{array}
$$

The domain of $(S \mathbin{⩤} f)$ is simply the domain of f with all elements of S removed

$$dom\ (S \mathbin{⩤} f) = (dom\ f) \setminus S$$

We could define domain subtraction in the following way, and it makes a useful law anyway

$$S \mathbin{⩤} R = (A \setminus S) \lhd R$$

where A is the type from which the members of S are drawn. Another very useful law is that a relation can always be expressed as the union of two others relative to a set S

$$R = (S \lhd R) \cup (S \mathbin{⩤} R)$$

Exercise 7.15 *Prove the laws of domain subtraction.*

Range Subtraction

Again by symmetry, we have range subtraction.

$$
\begin{aligned}
& dir \mathbin{⩥} \{u_4, u_{19}\} \\
& \quad = \{1 \mapsto u_2, 3 \mapsto u_4, 4 \mapsto u_4, 5 \mapsto u_4, 10 \mapsto u_7\} \rhd \{u_4, u_{19}\} \\
& \quad = \{1 \mapsto u_2, 10 \mapsto u_7\}\ dir \rhd \{u_2, u_7\}
\end{aligned}
$$

$$
\begin{array}{|l}
\hline
[A, B] \\
\_ \mathbin{⩥} \_ : (A \leftrightarrow B) \times (\mathbb{P}\, B) \rightarrow (A \leftrightarrow B) \\
\hline
\forall R : A \leftrightarrow B; S : \mathbb{P}\, B \bullet \\
\quad R \mathbin{⩥} S = \{a : A; b : B \mid a\ R\ b \land b \notin S \bullet a \mapsto b\} \\
\hline
\end{array}
$$

The range of $(f \rhd S)$ is simply the range of f without any element from S

$$ran\ (f \mathbin{⩥} S) = (ran\ f) \setminus S$$

The following laws should already have occurred to the reader

$$R \mathbin{⩥} S = R \rhd (B \setminus S)$$
$$R = (R \rhd S) \cup (R \mathbin{⩥} S)$$

where B is the type from which S's elements are drawn.

Exercise 7.16 *Prove the laws of range subtraction.*

Inversion

We can give a formal definition of relational inversion that we have used intuitively up until now. When we invert a relation we merely reverse the order of the constituent maplets

$$[A, B]$$
$$\_^{-1} : (A \leftrightarrow B) \rightarrow (B \leftrightarrow A)$$
$$\forall R : A \leftrightarrow B \bullet$$
$$\quad R^{-1} = \{a : A; b : B \mid a\ R\ b \bullet b \mapsto a\}$$

Exercise 7.17 *Prove the following laws*

$$(R^{-1})^{-1} = R$$
$$dom\ (R^{-1}) = ran\ R$$
$$ran\ (R^{-1}) = dom\ R$$

Relational Composition

Here we present the formal definition of relational composition, an operator that we met in the last chapter.

$$[A, B, C]$$
$$\_ ⨾ \_ : (A \leftrightarrow B) \times (B \leftrightarrow C) \rightarrow (A \leftrightarrow C)$$
$$\forall R : A \leftrightarrow B; S : B \leftrightarrow C \bullet$$
$$\quad R ⨾ S = \{a : A; b : B; c : C \mid a\ R\ b \wedge b\ S\ c \bullet a \mapsto c\}$$

Exercise 7.18 *Prove the following laws*

$$Id_A \fatsemi R = R$$
$$R \fatsemi Id_B = R$$
$$(R \fatsemi S) \fatsemi T = R \fatsemi (S \fatsemi T)$$

Operations on Functions

Functional Overriding

If we want to construct a new function from two old ones f and g, we might try to unite them. The union of two functions is certainly a relation, but it is a function only if the domains of f and g are *disjoint* (in the sense described earlier), or if the elements common to both domains are mapped to consistent elements in the ranges of f and g. More formally, if f and g are functions, then their union is functional if and only if f and g produce the same results for all values in their common domains

$$\begin{array}{l} \forall f, g : X \twoheadrightarrow Y \bullet \\ \quad f \cup g \in X \twoheadrightarrow Y \Leftrightarrow \\ \qquad \forall x : X \mid x \in (dom\ f) \cap (dom\ g) \bullet f\ x = g\ x \end{array}$$

Thus, we can only unite two functions to form a third if there is no possibility of confusion. If f and g contained different results for the same domain element, then we could avoid confusion by saying that in the event of a conflict g, say, is to be preferred over f. We express this preference by

$$f \oplus g$$

pronounced "f overridden by g". For example

$$\begin{array}{l} dir \oplus \{2 \mapsto u_2, 3 \mapsto u_2\} \\ \quad = \{1 \mapsto u_2, 3 \mapsto u_4, 4 \mapsto u_4, 5 \mapsto u_4, 10 \mapsto u_7\} \oplus \\ \qquad \{2 \mapsto u_2, 3 \mapsto u_2\} \\ \quad = \{1 \mapsto u_2, 2 \mapsto u_2, 3 \mapsto u_2, 4 \mapsto u_4, 5 \mapsto u_4, 10 \mapsto u_7\} \end{array}$$

We can use *dom*, domain subtraction, and union to give a formal definition of overriding. The part of f that cannot be overridden by g is

$$(dom\ g) \lhd f$$

that is, f without any of the maplets that could cause confusion. We can now unite this with g knowing that there can be no conflict, and having given preference to g over f. The domains of $(dom\ g) \lhd f$ and g are clearly disjoint, since

$$dom\ ((dom\ g) \lhd f) = (dom\ f) \setminus (dom\ g)$$

Summarising our definition, we have

$$[A, B]$$
$$\_ \oplus \_ : (A \nrightarrow B) \times (A \nrightarrow B) \rightarrow (A \nrightarrow B)$$
$$\forall f, g : A \nrightarrow B \bullet$$
$$f \oplus g = ((dom\ g) \lhd f) \cup g$$

Functional overriding has some pleasing properties, such as being idempotent and associative, and having left and right units

$$f \oplus f = f$$
$$f \oplus (g \oplus h) = (f \oplus g) \oplus h$$
$$\{\} \oplus f = f$$
$$f \oplus \{\} = f$$

If f and g have disjoint domains, then

$$f \oplus g = f \cup g$$

Thus, overriding *sometimes* commutes

$$dom\ f \cap dom\ g = \{\} \Rightarrow f \oplus g = g \oplus f$$

Exercise 7.19 *Prove the laws of functional overriding.*

7.3 Fancy Functions

In this section we offer some fancy functions to interest the reader. The funny symbols

$$⤔ \quad ↣ \quad ⤀ \quad ↠ \quad ⤗ \quad ⤖$$

look a little weird at first, but the concepts that they capture are important. We give particular kinds of functions special symbols, so that when we write a function definition we automatically concentrate our minds on whether or not our function possesses any of these special properties. Also, when we see a definition that contains one of these arrows, we see something important straightaway.

Injections

An *injection* is a function where each member of the domain is mapped onto a *different* member of the range. There must be no *converging* arrows. Formally, f is an injection if

$$\begin{array}{l} \forall x_1, x_2 : dom\ f \bullet \\ \qquad f\ x_1 = f\ x_2 \Rightarrow x_1 = x_2 \end{array}$$

Recall our original example in this chapter

$$dir = \{1 \mapsto u_2, 3 \mapsto u_4, 4 \mapsto u_4, 5 \mapsto u_4, 10 \mapsto u_7\}$$

This is not an injection, since three elements (3, 4, 5) are mapped onto the same value (u_4). If we take the inverse of *dir*, we easily see that it is not a *function* (see Figure 7.2); this is equivalent to *dir* not being an injection. We can think of functionality and injectivity as being *dual* notions: a function f is injective whenever f^{-1} is a function.

Exercise 7.20 *There are fifteen subsets of dir which are partial injections, find them. Are there any subsets which are total injections?*

We define the set of all *partial injections* from A to B as

$$\begin{array}{l} A ⤔ B \ \widehat{=} \\ \qquad \{f : A ⇸ B \mid \forall x_1, x_2 : dom\ f \bullet f\ x_1 = f\ x_2 \Rightarrow x_1 = x_2\} \end{array}$$

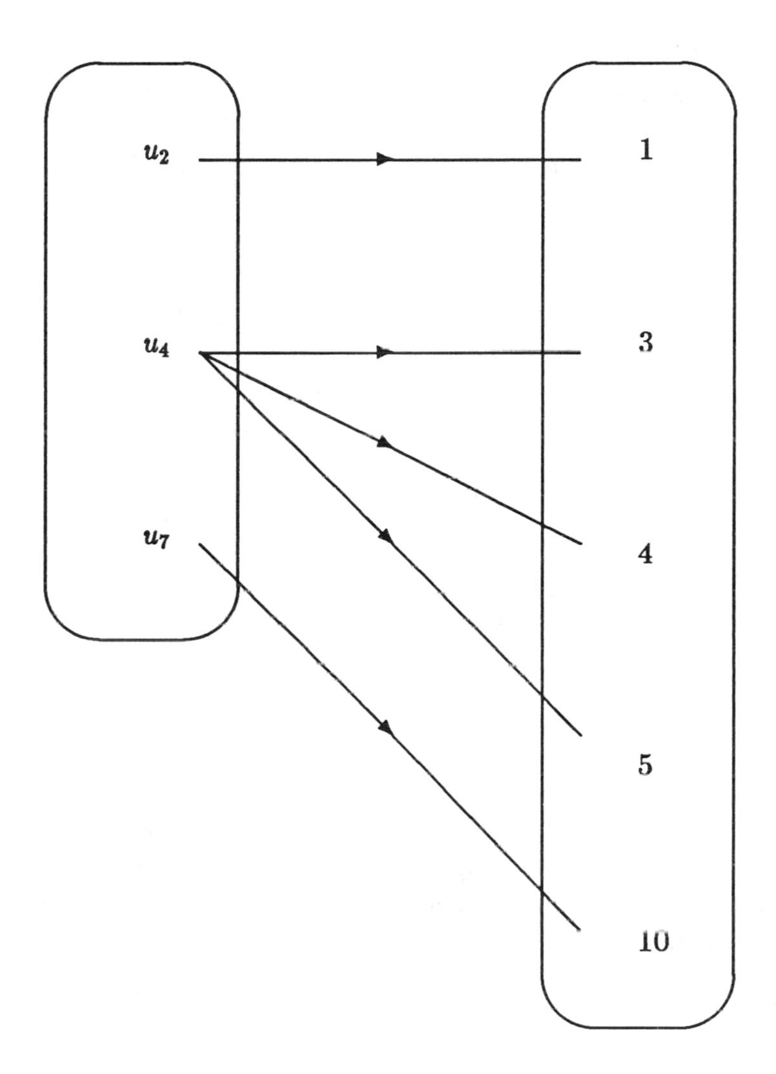

Figure 7.2: The relation dir^{-1}.

The *total injections* are all those total functions which are also injections

$$A \rightarrowtail B \triangleq (A \rightarrow B) \cap (A \mathrel{\rightarrowtail\!\!\!\!\!+} B)$$

Example 7.7 *Suppose that*

$$\begin{aligned} S &\triangleq \{a, b\} \\ T &\triangleq \{1, 2, 3\} \end{aligned}$$

then we can calculate the injections between S and T to be

$$\begin{aligned} S \mathrel{\rightarrowtail\!\!\!\!\!+} T = \{&\{\}, \{a \mapsto 1\}, \{a \mapsto 2\}, \{a \mapsto 3\}, \\ &\{b \mapsto 1\}, \{b \mapsto 2\}, \{b \mapsto 3\}, \\ &\{a \mapsto 1, b \mapsto 2\}, \{a \mapsto 1, b \mapsto 3\}, \\ &\{a \mapsto 2, b \mapsto 1\}, \{a \mapsto 2, b \mapsto 3\}, \\ &\{a \mapsto 3, b \mapsto 1\}, \{a \mapsto 3, b \mapsto 2\}\} \\ S \rightarrowtail T = \{&\{a \mapsto 1, b \mapsto 2\}, \{a \mapsto 1, b \mapsto 3\}, \\ &\{a \mapsto 2, b \mapsto 1\}, \{a \mapsto 2, b \mapsto 3\}, \\ &\{a \mapsto 3, b \mapsto 1\}, \{a \mapsto 3, b \mapsto 2\}\} \end{aligned}$$

Each member of $S \mathrel{\rightarrowtail\!\!\!\!\!+} T$ is a partial injection and each member of $S \rightarrowtail T$ is a total injection.

Surjections

A surjective function from A to B is one where the range is the whole of B. Recalling what we said about functionality and injectivity, it is worth noting here that totality and surjectivity are dual notions: a function f is surjective whenever f^{-1} is total. We define the set of all *partial surjections* from A to B as

$$A \mathrel{+\!\!\!\!\twoheadrightarrow} B \triangleq \{f : A \mathrel{+\!\!\!\!\rightarrow} B \mid ran\ f = B\}$$

The *total surjections* are all those total functions which are also surjections

$$A \twoheadrightarrow B \triangleq (A \rightarrow B) \cap (A \mathrel{+\!\!\!\!\twoheadrightarrow} B)$$

Exercise 7.21 *Calculate the partial and total surjections which are subsets of dir.*

Example 7.8 *If we try to calculate the partial surjective functions from S to T in Example 7.7, we can see that S doesn't contain enough values. In fact, there should be at least as many things in the domain of a surjection as there are in its range. Instead, we calculate the surjections from T to S*

$$
\begin{aligned}
T \mathrel{+\!\!\!\!\twoheadrightarrow} S = \{ & \{1 \mapsto a, 2 \mapsto b\}, \{1 \mapsto a, 3 \mapsto b\}, \{2 \mapsto a, 1 \mapsto b\}, \\
& \{2 \mapsto a, 3 \mapsto b\}, \{3 \mapsto a, 1 \mapsto b\}, \{3 \mapsto a, 2 \mapsto b\}, \\
& \{1 \mapsto a, 2 \mapsto a, 3 \mapsto b\}, \{1 \mapsto a, 2 \mapsto b, 3 \mapsto a\}, \\
& \{1 \mapsto b, 2 \mapsto a, 3 \mapsto a\}, \{1 \mapsto a, 2 \mapsto b, 3 \mapsto b\}, \\
& \{1 \mapsto b, 2 \mapsto a, 3 \mapsto b\}, \{1 \mapsto b, 2 \mapsto b, 3 \mapsto a\}\} \\
T \rightarrowtail S = \{ & \{1 \mapsto a, 2 \mapsto a, 3 \mapsto b\}, \{1 \mapsto a, 2 \mapsto b, 3 \mapsto a\}, \\
& \{1 \mapsto b, 2 \mapsto a, 3 \mapsto a\}, \{1 \mapsto a, 2 \mapsto b, 3 \mapsto b\}, \\
& \{1 \mapsto b, 2 \mapsto a, 3 \mapsto b\}, \{1 \mapsto b, 2 \mapsto b, 3 \mapsto a\}\}
\end{aligned}
$$

Bijections

We define the set of all *partial bijections* from A to B to be those partial functions that are both injections and surjections

$$A \mathrel{\rightarrowtail\!\!\!\!\!\!+\!\!\!\twoheadrightarrow} B \;\widehat{=}\; (A \mathrel{\rightarrowtail\!\!\!\!\!\!+} B) \cap (A \mathrel{+\!\!\!\!\twoheadrightarrow} B)$$

Finally, we define the set of all *total bijections* from A to B to be those partial bijections that are total

$$A \mathrel{\rightarrowtail\!\!\!\!\twoheadrightarrow} B \;\widehat{=}\; (A \mathrel{\rightarrowtail\!\!\!\!\!\!+\!\!\!\twoheadrightarrow} B) \cap (A \rightarrow B)$$

Exercise 7.22 *Calculate the partial and total bijections which are subsets of dir.*

Example 7.9 *There are no bijections of either sort between S and T from Example 7.7, and consequently none between T and S.*

The Natural Numbers

In Chapter 5 we introduced the natural numbers intuitively as the set $\mathbb{N}$, and then used them freely. We have now reached the stage where we can give a more precise formulation: $\mathbb{N}$ contains the element *zero*; and there is a total bijection from $\mathbb{N}$ to $\mathbb{N} \setminus \{zero\}$, called the *successor* function; *zero* is a distinguished natural number which is not the successor of any number

$$zero : \mathbb{N}$$
$$succ : \mathbb{N} \rightarrowtail\!\!\!\!\twoheadrightarrow (\mathbb{N} \setminus \{zero\})$$

Thus, *zero* is in $\mathbb{N}$, and *succ n* is in $\mathbb{N}$ whenever n is in $\mathbb{N}$. We use particular symbols to denote the natural numbers, such as 0 for *zero*, 1 for *succ zero*, 2 for *succ* (*succ zero*), and so on.

Exercise 7.23 *Not only have we used the natural numbers informally, we have also used many of the operators on the naturals without definition.*

1. *Define the usual operations of addition, subtraction, and multiplication on the natural numbers.*
2. *Write down and prove correct some laws governing these operations.*
3. *Define the ordering relation on the natural numbers* $m \leq n$.

Exercise 7.24 *If S is any nonempty collection of natural numbers, prove that S has a* least *element. This is known as the* Well-Ordering Principle. *[Hint: Start by considering, for every n, the set $1\,..\,n$ of natural numbers not in S.]*

Finite Sets

Subrange

We use the expression

$$4\,..\,8$$

(pronounced 4 *up to* 8) as a convenient shorthand for the set

$$\{4, 5, 6, 7, 8\}$$

Other examples are

$$6 \mathinner{..} 2 = \{\}$$
$$42 \mathinner{..} 42 = \{42\}$$

The definition makes this precise

$$\begin{array}{|l} \_ \mathinner{..} \_ : \mathbb{N} \times \mathbb{N} \to \mathbb{P}\,\mathbb{N} \\ \hline \forall m, n : \mathbb{N} \bullet m \mathinner{..} n = \{i : \mathbb{N} \mid m \leq i \leq n\} \end{array}$$

Cardinality

A *finite* set is one whose elements are countable up to some natural number.

Example 7.10 *We have just had some obvious examples of finite sets*

$$4 \mathinner{..} 8,\ 6 \mathinner{..} 2,\ 42 \mathinner{..} 42$$

These three sets have 5, 0, *and* 1 *elements, respectively. In general, the set* $n \mathinner{..} m$ *is finite, with* $m - n + 1$ *elements if it is nonempty, or* 0 *elements if it is empty.*

We can extend our example to describe all possible finite sets: a set S is *finite* if it is possible to enumerate its elements up to some natural number. Our task may be described as finding some *total bijection* f between the set S and an initial segment of the natural numbers $1 \mathinner{..} n$. We know that $1 \mathinner{..} n$ is finite—it has n elements. The first element of S is simply $f\ 1$, the second $f\ 2$, ..., and the last $f\ n$. Consider the set

$$IBMproject = \{ib, jim, steve, ib\}$$

The cardinality of *IBMproject* is obviously 3, since we already know that duplicate elements in a set may be disregarded. However, if we didn't know this fact it wouldn't matter, since we cannot produce a bijection from an initial segment of the naturals into *IBMproject* which has each occurrence of *ib* being mapped to by distinct natural numbers: that would violate the *injectivity* of our bijection. Thus, we cannot count any elements more than once. That we must count each element at least once is guaranteed by the *surjectivity* of our bijection. We can summarise our definition formally

$$\mathbb{F}\, X \mathrel{\widehat{=}} \{S : \mathbb{P}\, X \mid \exists\, n : \mathbb{N};\, f : 1 \mathinner{..} n \rightarrowtail\!\!\!\!\twoheadrightarrow S \bullet \mathit{true}\}$$

Example 7.11 *Referring to our previous examples,*

1. $4 \mathinner{..} 8$ *is a finite set, since we can find the bijection*

$$\{i : 1 \mathinner{..} 5 \bullet i \mapsto i + 3\}$$

2. $6 \mathinner{..} 2$ *is a finite set, since we can find the bijection*

$$\{i : 1 \mathinner{..} 0 \bullet i \mapsto i\}$$

3. $42 \mathinner{..} 42$ *is a finite set, since we can find the bijection*

$$\{i : 1 \mathinner{..} 1 \bullet i \mapsto i\}$$

In the definition of finite sets, the number n in the set $1 \mathinner{..} n$ is called the *cardinality* of the set. Obviously, only finite sets have a cardinality. We shall find many uses for the cardinality of a set S, and we shall denote it by $\#S$

$$\begin{array}{|l}
[X] \\
\# : \mathbb{F}\, X \rightarrow \mathbb{N} \\
\hline
\forall\, S : \mathbb{F}\, X \bullet \\
\quad \exists\, f : 1 \mathinner{..} \#S \rightarrowtail\!\!\!\!\twoheadrightarrow S \bullet \mathit{true}
\end{array}$$

Example 7.12 *From our earlier example*

$$\#(4 \mathinner{..} 8) = 5, \#(6 \mathinner{..} 2) = 0, \#(42 \mathinner{..} 42) = 1$$

Although there may be many bijections that we can construct between an initial segment of the natural numbers and a finite set S, the cardinality of S is uniquely defined. A proof of this fact is quite rightly outside the scope of this book: we are trying to introduce just enough mathematics to be able to describe software systems, not cover all of set theory. However, readers whose interest has been stimulated in this and other foundational matters are referred to the literature[7].

Finite Functions

In the next chapter we shall find it convenient to use functions which are finite. We shall denote the set of finite partial functions from A to B by $A \twoheadrightarrow\!\!\!\!\!\!\!\!+ \; B$. The description presents no problems

$$A \;\text{-}\!\!\!\twoheadrightarrow B \;\widehat{=}\; \{f : A \nrightarrow B \mid dom\, f \in \mathbb{F}\, A\}$$

The set of finite injections is

$$A \rightarrowtail\!\!\!\twoheadrightarrow B \;\widehat{=}\; A \;\text{-}\!\!\!\twoheadrightarrow B \cap A \rightarrowtail\!\!\!\!\!\!+\; B$$

7.4 Example: Storage Allocator

We now present a small case study which uses several of the ideas introduced in this chapter.

Requesting and Releasing Blocks

In an operating system there is a program which manages blocks of storage (drawn from the set B) which users (drawn from the set U) may want to access.

There are n consecutively numbered blocks

[7]For example, P.R. Halmos, *Naïve Set Theory*, Undergraduate Texts in Mathematics, New York: Springer-Verlag (1970), *104 pp*, or S. Hayden & J.F. Kennison, *Zermelo-Frænkel Set Theory*, Merrill Mathematics Series, Columbus Ohio: Merrill (1968), *164 pp*.

$$\begin{array}{|l} n : \mathbb{N} \\ B : \mathbb{P}\,\mathbb{N} \\ \hline B = 1 \mathinner{\ldotp\ldotp} n \end{array}$$

The storage manager maintains a directory of which blocks are owned by which users. We call this information structure *dir*, and we would like it to have various properties as a relation

1. No block is owned by more than one user.

2. A user may own more than one block.

3. Some blocks might not be owned at all.

4. Some users might not own any blocks.

We can express these properties more formally

1. *dir* is *functional.*

2. *dir* need not be an *injection.*

3. *dir* may be *partial.*

4. *dir* need not be a *surjection.*

So it is enough to say that *dir* is a partial function from B to U

$$dir : B \rightarrowtail U$$

We also need a record of which blocks are *free*, that is, not owned by anyone. This is simply a subset of B

$$free : \mathbb{P}\, B$$

Of course, it must be that the free blocks are all the ones not being used. We can specify this

$$free = B \setminus (dom\ dir)$$

We call this predicate the *state invariant*: it must always be true. A function

$$dir : B \rightarrowtail U$$

and a set

$$free : \mathbb{P}\ B$$

do not constitute a possible state of the storage manager unless they satisfy the state invariant.

The state invariant, together with the declarations of *dir* and *free*, form a schema which we shall call *SM*.

```
┌─ SM ──────────────────────────
│ dir : B ⇸ U
│ free : ℙ B
├───────────
│ free = B \ (dom dir)
└───────────────────────────────
```

We shall often want to refer to the state before an operation, and the state after it. Our state transitions—the operations—can then be described by the relation between these two states, as we have seen before. Define

```
┌─ ΔSM ─────────────────────────
│ SM
│ SM'
└───────────────────────────────
```

ΔSM then is a shorthand for the longer text

```
┌─ ΔSM ─────────────────────────
│ dir, dir' : B ⇸ U
│ free, free' : ℙ B
├───────────
│ free = B \ (dom dir)
│ free' = B \ (dom dir')
└───────────────────────────────
```

We use the symbol Δ in the name of this schema to evoke the notion of change. Whenever we see ΔSM in the declaration part of a schema, we know that it is describing an operation that changes the state.

The initial state of our system has no blocks allocated to any user, and every block free

$$\begin{array}{|l}
\multicolumn{1}{l}{__ \textit{InitSM} ________________} \\
SM' \\
\hline
dir' = \{\} \\
free' = B \\
\hline
\end{array}$$

We must ensure that such an initial state really exists; a simple check suffices:

$$\begin{aligned}
free' &= B \\
&= B \setminus \{\} \\
&= B \setminus (\mathrm{dom}\ dir')
\end{aligned}$$

Thus dir' and $free'$ satisfy the invariant.

We decorate inputs with a final question mark, and outputs with a final exclamation mark. The following operation corresponds to user $u?$ requesting and being allocated a free block $b!$, and receiving the response that the operation was performed satisfactorily. There ought to be some block that can be allocated

$$free \neq \{\}$$

The block being allocated really was free

$$b! \in free$$

After the operation, it is no longer free

$$free' = free \setminus \{b!\}$$

and has been allocated to the user

$$dir' = dir \cup \{b! \mapsto u?\}$$

The state invariant must hold both before and after the operation.

$$
\begin{array}{|l}
\hline
\textit{Request}_0 \\
\Delta SM \\
u? : U \\
b! : B \\
r! : Report \\
\hline
free \neq \{\} \\
b! \in free \\
free' = free \setminus \{b!\} \\
dir' = dir \cup \{b! \mapsto u?\} \\
r! = okay \\
\hline
\end{array}
$$

Although we have specified that the state invariant is preserved, how can we tell that this operation really does just that? We must calculate the *precondition*: the set of states from which it is guaranteed that a final state can be reached. The precondition of an operation is analogous to the domain of a relation. So, for each state in the precondition, we must show that there exists a final state related to it by the operation. The following theorem shows how to calculate the precondition for $Request_0$.

Theorem 7.1 *The precondition for* $Request_0$ *is just* $free \neq \{\}$. *Let* $PreRequest_0$ *denote the precondition for* $Request_0$*, then we must show that* $PreRequest_0$ *is simply*

$$
\begin{array}{|l}
\hline
\textit{PreRequest}_0 \\
SM \\
u? : U \\
\hline
free \neq \{\} \\
\hline
\end{array}
$$

Proof: First we expand $PreRequest_0$ *as the schema containing the state* SM *constrained so that a final state can be reached. Here the final state is existentially quantified: it must satisfy the predicates of* $Request_0$*; and also the state invariant*

PreRequest$_0$

$$
\begin{array}{|l}
SM \\
u? : U \\
\hline
\exists\, dir' : B \pfun U;\ free' : \mathbb{P}\, B;\ b! : B;\ r! : Report \bullet \\
\qquad free \neq \{\} \\
\qquad b! \notin free \\
\qquad free' = free \setminus \{b!\} \\
\qquad dir' = dir \cup \{b! \mapsto u?\} \\
\qquad r! = okay \\
\qquad free' = B \setminus (\mathrm{dom}\ dir') \\
\hline
\end{array}
$$

Next, we can eliminate dir', $free'$, and $r!$ from the quantified predicate; notice that we retain two proof obligations

1. *dir' is not just a relation between B and U, but a* partial function.

2. *$free'$ is not just a subset of $\mathbb{N}$, but a subset of B.*

PreRequest$_0$

$$
\begin{array}{|l}
SM \\
u? : U \\
\hline
\exists\, b! : B \bullet \\
\qquad dir \cup \{b! \mapsto u?\} \in B \pfun U \\
\qquad free \setminus \{b!\} \subseteq B \\
\qquad free \neq \{\} \\
\qquad b! \in free \\
\qquad free \setminus \{b!\} = B \setminus (\mathrm{dom}\ (dir \cup \{b! \mapsto u?\})) \\
\hline
\end{array}
$$

It is fairly easy to dispose of the first of the quantified conjuncts, since it happens to be equivalent to $b! \in free$

$$
\begin{array}{rcl}
b! \in free & \Leftrightarrow & b! \in B \setminus (\mathrm{dom}\ dir) \\
 & \Leftrightarrow & b! \notin \mathrm{dom}\ dir \\
 & \Leftrightarrow & dir \in b \pfun U \wedge b! \notin \mathrm{dom}\ dir \\
 & \Leftrightarrow & dir \cup \{b! \mapsto u?\} \in B \pfun U
\end{array}
$$

The second is even easier, depending as it does on the fact that

$$free \in \mathbb{P}\, B$$

and the transitivity of $\subseteq$

$$free \setminus \{b!\} \subseteq free \subseteq B$$

Lastly, consider

$$\begin{aligned} B \setminus (dom\ (dir \cup \{b! \mapsto u?\})) &= B \setminus ((dom\ dir) \cup \{b!\}) \\ &= (B \setminus (dom\ dir)) \setminus \{b!\} \\ &= free \setminus \{b!\} \end{aligned}$$

Using these results, we can rewrite our schema as

$$\begin{array}{|l} \text{\_ } PreRequest_0 \text{ \_\_\_\_\_\_\_\_\_\_\_\_\_\_\_\_\_\_\_\_} \\ SM \\ u? : U \\ \hline \exists\, b! : B \bullet \\ \quad free \neq \{\} \\ \quad b! \in free \\ \hline \end{array}$$

But of course, if free is not empty, then we can always find an element in it. So we have

$$\begin{array}{|l} \text{\_ } PreRequest_0 \text{ \_\_\_\_\_\_\_\_\_\_\_\_\_\_\_\_\_\_\_\_} \\ SM \\ u? : U \\ \hline free \neq \{\} \\ \hline \end{array}$$

QED.

So, the operation of requesting a block from the storage manager is a partial one: $Request_0$ cannot proceed successfully if $free = \{\}$. If we try to apply the operation when no free block can be allocated, then we are in considerable trouble! To prevent this unsatisfactory state of affairs, we must specify what should happen in the *unsuccessful* case. We shall define the error case using a schema that ensures that the state does not change

$$
\begin{array}{|l}
\hline
\textit{RequestErr} \\
\Xi SM \\
r! : Report \\
\hline
free = \{\} \\
r! = fail \\
\hline
\end{array}
$$

where ΞSM is a schema that permits a state change (it includes the schema ΔSM), but describes a transition that leads back to the same state

$$
\begin{array}{|l}
\hline
\Xi SM \\
\Delta SM \\
\hline
free = free' \\
dir = dir' \\
\hline
\end{array}
$$

Whenever we see a schema that includes ΞSM in its declaration part, we know that it describes an operation that leaves the state unchanged Calculating the precondition for *RequestErr* is much simpler than its successful counterpart: it is that

$$free = \{\}$$

Now our complete request operation—which is total—is defined as

$$Request \mathrel{\widehat=} Request_0 \vee RequestErr$$

The disjunction between the two schemas is used to structure the operation: either the successful case holds, or the error case does. We can destroy the structure and retrieve the entire description, by merging the declarations and disjoining the predicates

$$
\begin{array}{|l}
\hline
\textit{RequestBlock} \\
\Delta SM \\
u? : U \\
b! : \mathbb{N} \\
r! : Report \\
\hline
(\quad free \neq \{\} \\
\quad\quad b! \in free \\
\quad\quad free' = free \setminus \{b!\} \\
\quad\quad dir' = dir \cup \{b! \mapsto u?\} \\
\quad\quad r! = okay \quad) \\
\vee \\
(\quad free = \{\} \\
\quad\quad r! = fail \\
\quad\quad free = free' \\
\quad\quad dir = dir' \quad) \\
\hline
\end{array}
$$

Notice that in merging the two schemas, the requirement that

$$b! \in B$$

occurs only in the first disjunct. Thus, if this operation fails, we cannot even assert that the output $b!$ refers to an element of B; it might be any number.

The complementary operation is that of releasing a block: *Release*. Again, we describe the successful operation, and then consider the erroneous cases. When *Release* is invoked, the user and block must be given, and a report on the success of the operation is returned. In order for *Release* to be successful, the block $b?$ must indeed be owned by the user $u?$

$$(b? \mapsto u?) \in dir$$

Releasing a block makes it free again

$$free' = free \cup \{b?\}$$

which means that nobody owns it

$$dir' = \{b?\} \lhd dir$$

The successful part of the operation is

$$
\begin{array}{l}
\textit{Release}_0 \\
\hline
\Delta SM \\
u? : U \\
b? : B \\
r! : \textit{Report} \\
\hline
(b? \mapsto u?) \in \textit{dir} \\
\textit{free}' = \textit{free} \cup \{b?\} \\
\textit{dir}' = \{b?\} \lhd \textit{dir} \\
r! = \textit{okay} \\
\hline
\end{array}
$$

We can conceive of two cases where *Release* may be incorrectly invoked: when the block in question is already free; and when the block is owned by a different user. The two cases are described in the schemas *RelFreeErr* and *RelOwnerErr*, respectively

$$
\begin{array}{l}
\textit{RelFreeErr} \\
\hline
\Xi SM \\
u? : U \\
b? : B \\
r! : \textit{Report} \\
\hline
b? \in \textit{free} \\
r! = \textit{BlockFree} \\
\hline
\end{array}
$$

$$
\begin{array}{l}
\textit{RelOwnerErr} \\
\hline
\Xi SM \\
u? : U \\
b? : B \\
r! : \textit{Report} \\
\hline
b? \in \textit{dom dir} \\
\textit{dir}\ b? \neq u? \\
r! = \textit{NotOwner} \\
\hline
\end{array}
$$

Operation	Input	Output	Precondition
InitSM			*true*
$Request_0$	$u? : U$	$b! : B$ $r! : Report$	$free \neq \{\}$
RequestErr		$r! : Report$	$free = \{\}$
$Release_0$	$u? : U$ $b? : B$	$r! : Report$	$(b? \mapsto u?) \in dir$
RelFreeErr	$u? : U$ $b? : B$	$r! : Report$	$b? \in free$
RelOwnerErr	$u? : U$ $b? : B$	$r! : Report$	$b? \in \mathrm{dom}\ dir$ $dir\ b? \neq u?$

Figure 7.3: Preconditions for the Storage Allocation Operations

The total operation is the disjunction of these three schemas

$$Release \mathrel{\widehat{=}} Release_0 \lor RelFreeErr \lor RelOwnerErr$$

To summarise the operations, we present their preconditions in Figure 7.3

Requesting Contiguous Sets of Blocks

In a real operating system, users would not be content to have just a block allocated at a time—they tend to want to have a number of blocks allocated together. Moreover, they tend to want to have a *contiguous* set of blocks, that is, a set of adjoining blocks. How might our operation look with this modified requirement? Let's define what we mean by *contiguity.* A set of numbers s is contiguous if s contains a lowest number—say l—and a highest number—say h—and every number between l and h is in s

$$\exists\, l, h : \mathbb{N} \mid l \in s \land h \in s \bullet s = l \mathinner{..} h$$

But, what we are requiring is that s is one of the sets constructed by the function $(\_ \ldots \_)$. Therefore, a more succinct definition would be[8]

$$s \in \mathit{ran}\ (\_ \ldots \_)$$

We call our operation to request a contiguous set of blocks *ReqStore*. It requires as input a user $u?$ and the number of blocks being requested $n?$. It offers in return a set of blocks $b!$. The successful case works as follows. The set of blocks output must be of the requisite size

$$\#b! = n?$$

$b!$ must be a contiguous set

$$b! \in \mathit{ran}\ (\_ \ldots \_)$$

and of course they must all be free

$$b! \subseteq \mathit{free}$$

After the operation all the blocks in $b!$ are booked out to $u?$

$$\mathit{dir}' = \mathit{dir} \cup (b! \times \{u?\})$$

In this predicate, the expression

$$b! \times \{u?\}$$

denotes the function that maps every element of $b!$ to $u?$. After the operation none of the members of $b!$ are free any more

$$\mathit{free}' = \mathit{free} \setminus b!$$

Putting all this together gives us the description of the successful operation

---

[8] This definition is an example of something called a "Sørensen shortie", and was suggested by J.M. Spivey.

$$
\begin{array}{|l}
\hline
\textit{ReqStore}_0 \\
\Delta SM \\
u? : U \\
n? : B \\
b! : \mathbb{P}\, B \\
r! : Report \\
\hline
\#b! = n? \\
b! \in \mathrm{ran}\ (\_ \mathinner{\ldotp\ldotp} \_) \\
b! \subseteq free \\
dir' = dir \cup (b! \times \{u?\}) \\
free' = free \setminus b! \\
r! = okay \\
\hline
\end{array}
$$

In the implementation of such a storage manager, we will have eventually to decide upon an *allocation policy*. If we want $n?$ consecutive blocks of store, then we must locate a contiguous set of at least $n?$ in size. There may be several sets which satisfy this criterion, so which should we choose? There are two recommended answers to this problem[9]. We can choose the "best fit" method or the "first fit" method. The former chooses a contiguous set of size m, where m is the smallest value present which is $n?$ or more; $n?$ of these are then allocated. The latter simply chooses the first contiguous set encountered that has at least $n?$ blocks. We describe the first fit method. The set s of the *start points* of all contiguous sets of blocks which could satisfy a request for $n?$ blocks is

$$s = \{l, h : B \mid l \mathinner{\ldotp\ldotp} h \subseteq free \wedge l - 1 \notin free \wedge h - l + 1 \geq n? \bullet l\}$$

The set that we want starts with the minimum value of s, and goes on for $n?$ blocks

$$b! = (min\ s) \mathinner{\ldotp\ldotp} (min\ s) + n? - 1$$

We describe our method as a relation between SM and the output

[9] See D.E. Knuth, *The Art of Computer Programming*, Vol 1, "Fundamental Algorithms", Addison-Wesley, 1968, for a description of both methods and a comparison of their relative merits.

$b!$[10]

$$
\begin{array}{|l}
\textit{FirstFit} \\
\hline
SM \\
b! : \mathbb{P}\, B \\
\hline
\exists s : \mathbb{P}\, B \bullet \\
\quad s = \{l, h : B \mid l \mathinner{..} h \subseteq free \land \\
\qquad\qquad l - 1 \notin free \land h - l + 1 \geq n? \bullet l\} \\
\quad b! = (min\ s) \mathinner{..} (min\ s) + n? - 1 \\
\hline
\end{array}
$$

The quantification in the predicate part serves to introduce s as a local variable. This is handy because of the need to refer to its value twice, and although mathematics comes for free, it makes the expression easier to read.

We can now add the requirement that the contiguous set of blocks selected is the first that could satisfy the request by conjoining our two schemas

$$ReqStoreFF_0 \mathrel{\widehat=} ReqStore_0 \land FirstFit$$

7.5 Summary

This chapter has introduced the notion of a function, and also provided most of the things that we have found useful in specification work: ways of building and manipulating sets, relations, and functions. This development of a rattle-bag of mathematical operators follows that of the Z basic library[11]. We have illustrated the use of functions by giving the specification of the interface to a storage allocator. The style of specification used is typical of the *model-oriented approach*.

[10] Actually, it is only necessary to define *FirstFit* as a relation between *free* and $b!$, but we have put *SM* in the declaration part of the schema to emphasise the connection with the request operation.

[11] J.M. Spivey, *The Z Reference Manual*, Prentice-Hall, 1988.

Chapter 8

Sequences

8.1 Introduction

The process of abstraction involves finding a description of a particular object that contains just enough information—not too much. The mathematical data types that we have introduced so far allow us to record collections of objects as sets, and the relationships between them as relations and functions. Sometimes it is important to place these objects in some sort of order, they may be sorted according to some key for example. In order to describe a collection of objects whose order *is* important, we introduce a new data type: the *sequence*[1]. In fact, we already have enough notation to describe ordered sets, so we shall build on this.

Suppose that we have three tasks t, u, and v, drawn from the set of all tasks T, which request pages of backing store from an operating system. If they make their requests in the order t, u, and then v, the state of the queue of tasks waiting for pages may be modelled as

$$bsq = \{1 \mapsto t, 2 \mapsto u, 3 \mapsto v\}$$

That is, t is first, u is second, and v is third. Clearly, bsq is a function from $\mathbb{N}$ to T. But it is more than that: the *domain* of bsq must start

[1] In the literature, the reader may find sequences variously called strings, lists, or traces, and $seq[X]$ is often written as X^*, X-*list*, or even list of X

at 1, we must know who is first, and mustn't have any holes in it, we must know all its entries.

This motivates our model of sequences of objects of type T: they are functions from $\mathbb{N}$ to T whose domain is a segment $1 \mathinner{..} n$, for some $n \in \mathbb{N}$. By modelling sequences in this way, we can build upon existing knowledge of functions, which are built upon relations, which are built upon sets.

We define the set of all sequences over a set X

$$seq[X] \triangleq \{f : \mathbb{N} \twoheadrightarrow\!\!\!\!\!\!+ \; X \mid \operatorname{dom} f = 1 \mathinner{..} \#f\}$$

A consequence of our definition is that our sequences are finite in length. For our purposes, this is adequate and avoids some irritating problems: for example, the length of a sequence is always defined, and there is always an end to every sequence. However, some applications of mathematics genuinely require the use of infinite sequences, finite ones just won't quite do. For instance, the description of the fairness of scheduling in an operating system requires the consideration of infinite sequences of requests and services, fairness being a property in the limit. Of course, since our finite sequences are built upon functions, and not in any sense primitive mathematical objects, the reader could equally well construct a theory of infinite sequences, if that were appropriate to the problem in hand.

Two Simple Operations and Some Notation

We can *index* our sequences merely by using them as functions. Thus, the ith element of a sequence s is

$$s\ i$$

provided, of course, that $i \in \operatorname{dom} s$. Thus,

$$bsq\ 2 = u$$

We can determine the *length* of a sequence s by finding its cardinality as a set, since there is a maplet in s for each element of the sequence

$$\#s$$

The length of our backing store queue is just the number of pairs in the mapping *bsq*

$$\# bsq = 3$$

As we shall see later, sequences are powerful aids to specification, and so they have a notation of their own and a collection of useful operators and functions. We shall write *bsq* as

$$bsq = \langle t, u, v \rangle$$

Here, we are using the obvious notation of displaying the order of the elements of a sequence, suppressing the formal clutter that we had before. In general, the shorthand is

$$\langle s_1, s_2, \ldots, s_n \rangle \quad \text{for} \quad \{1 \mapsto s_1, 2 \mapsto s_2, \ldots, n \mapsto s_n\}$$

with the *empty sequence* denoted $\langle \rangle$, rather than $\{\}$.

8.2 Operations on Sequences

The benefit of the new notation is most obvious when we consider some simple operations on sequences.

Catenation

Catenation constructs a new sequence from two others: s *catenate* t is simply formed by taking the elements of s followed by the elements of t. We write this new sequence as[2]

$$s \frown t$$

For illustration

$$\begin{aligned} \langle t, u, v \rangle \frown \langle w \rangle &= \langle t, u, v, w \rangle \\ \langle 0 \rangle \frown \langle 1 \rangle &= \langle 0, 1 \rangle \\ \langle \rangle \frown \langle v, t \rangle &= \langle v, t \rangle \end{aligned}$$

[2] In the literature, catenation is also written as $s\hat{}t$, $s \| t$, $append(s, t)$, or $cons(s, t)$.

The definition of catenation using our model is complicated only by the need to adjust the indices of the second sequence's elements. If s and t are sequences, of the same type, then we can index $s \frown t$ as follows

$$(s \frown t)\ i = \begin{cases} s\ i & \text{if } i \in 1 \mathinner{\ldotp\ldotp} \#s \\ t\ (i - \#s) & \text{if } i \in (\#s + 1) \mathinner{\ldotp\ldotp} (\#s + \#t) \end{cases}$$

This may be expressed as the union of two functions, s and

$$(\lambda\, n : \mathbb{N} \mid n > \#s \bullet n - \#s) \mathbin{;} t$$

This last function deals with the indices that lie in the set

$$(\#s + 1) \mathinner{\ldotp\ldotp} (\#s + \#t)$$

The lambda expression does the necessary jiggery-pokery, decrementing indices by $\#s$, before t is indexed. It should be clear that the result of uniting these two functions is itself a function, and moreover a *sequence*. Demonstrating this is a proof obligation. Thus we can give a formal definition of catenation

$$[X]$$

$$\_ \frown \_ : seq[X] \times seq[X] \to seq[X]$$

$$\forall\, s, t : seq[X] \bullet$$
$$\quad s \frown t = s \cup ((\lambda\, n : \mathbb{N} \mid n > \#s \bullet n - \#s) \mathbin{;} t)$$

Catenation is associative and has $\langle\rangle$ as its identity

$$s \frown (t \frown u) = (s \frown t) \frown u$$
$$s \frown \langle\rangle = \langle\rangle \frown s = s$$

The preceding discussion should convince the reader that

$$\#(s \frown t) = \#s + \#t$$

but if not...

A function on sequences $f : seq[X] \twoheadrightarrow seq[X]$ is said to be ***strict*** if it preserves the empty sequence, and ***distributive*** if it distributes through catenation

$$f \langle\rangle = \langle\rangle$$
$$f\ (s \frown t) = (f\ s) \frown (f\ t)$$

We shall see several examples of strict and distributive functions later in the chapter.

Example: A Backing Store

Consider our backing store example in a little more detail. Tasks make some request for pages of backing store; sometime later the operating system considers the request; eventually the pages get allocated. Tasks are served strictly in the order in which the requests were made. We shall model this using three sequences

inq for waiting tasks

q for tasks currently being considered

$outq$ for tasks whose requests have been completed

We describe our state as follows:

$$\begin{array}{l} \underline{\ BSQ\ } \\ inq, q, outq, bsq : seq[X] \\ \hline bsq = outq \frown q \frown inq \end{array}$$

As you can see, we have added a fourth item to our state: the sequence bsq, the backing store queue, which always consists of the other three sequences catenated together. A task will arrive at the back of bsq, and work its way towards the head of bsq. We define two operations which enqueue and dequeue tasks. ENQ is an operation on the state BSQ, and takes as its input the name of a task $t?$, which must not already be in any of the backing store queues

$$t? \notin ran\ bsq$$

This task name is put at the end of the queue of waiting tasks

$$inq' = inq \frown \langle t? \rangle$$

Enqueuing a task changes neither the current queue, nor the completed queue

$$q' = q \land outq' = outq$$

The entire operation is described as

$$\begin{array}{|l} \hline \textit{ENQ} \\ \Delta BSQ \\ t? : T \\ \hline t? \notin \operatorname{ran} bsq \\ inq' = inq \frown \langle t? \rangle \\ q' = q \\ outq' = outq \\ \hline \end{array}$$

DEQ is again an operation on the state *BSQ*, returning as an output $t!$, the name of the task which it removed from the front of the completion queue

$$outq = \langle t! \rangle \frown outq'$$

The other two queues are unaffected by this operation

$$inq' = inq \land q' = q$$

DEQ in full is

$$\begin{array}{|l} \hline \textit{DEQ} \\ \Delta BSQ \\ t! : T \\ \hline outq = \langle t! \rangle \frown outq' \\ inq' = inq \\ q' = q \\ \hline \end{array}$$

Finally we describe the two operations that transfer tasks from the waiting queue to the current queue, and from the current queue to the completed queue. The first of these two, *Start*, transfers a task name from the waiting queue to the current queue. The relative order of tasks in the various queues is left unchanged by this operation

$$bsq' = bsq$$

The completed queue doesn't change

$$outq' = outq$$

The current queue grows by one element

$$\#q' = \#q + 1$$

Putting all this together we get

Start

$$\Delta BSQ$$

$$outq' = outq$$
$$bsq' = bsq$$
$$\#q' = \#q + 1$$

The complementary operation *Complete* is very similar

Complete

$$\Delta BSQ$$

$$inq' = inq$$
$$bsq' = bsq$$
$$\#outq' = \#outq + 1$$

The description of *Start* and *Complete* suggests an implementation for the backing store queues: we could use just one array and have some pointers delimiting the queues; this would reflect the predicates that say that $\#q$ and $\#outq$ change. We could even make bsq into a circular queue—this might be convenient in a low-level operating system, since we could pre-allocate storage for the queue. Such a queue is sometimes called a ***barrel queue***. The preconditions for the various operations on the backing store queues are recorded in Figure 8.1.

Operation	Input	Output	Precondition
ENQ	$t? : T$		$t? \notin \text{ran } bsq$
DEQ		$t! : T$	$outq \neq \langle\rangle$
$Start$			$inq \neq \langle\rangle$
$Complete$			$q \neq \langle\rangle$

Figure 8.1: Preconditions for the Backing Store Operations

Exercise 8.1 *Prove that each operation really does have the precondition stated in Figure 8.1.*

Heads or Tails?

When we want to remove elements from sequences in operations we often want to refer to the first element of a sequence—its *head*, and sometimes to the decapitated sequence—its *tail*.

$$\begin{aligned} head\ \langle t, u, v\rangle &= t \\ tail\ \langle t, u, v\rangle &= \langle u, v\rangle \end{aligned}$$

We can define *head* as follows

$$\begin{array}{|l} [X] \\ \hline head : seq[X] \rightarrow\!\!\!\!\!\rightarrow X \\ \hline \forall s : seq[X] \bullet head\ s = s\ 1 \\ \hline \end{array}$$

Notice two wholly obvious, but nonetheless important, points: that *head* is *partial*—an empty sequence has no head—and that the head of a sequence of X's is an X. Beheading discards a level of sequencing. The fact that *head* is not defined for the empty sequence means that there is always a trap for the unwary: we must be careful to avoid it.

The complementary part of a sequence s is known as the tail of s which is a *sequence* containing every element of s, except the first. Of course, when we index the tail of s at 1 we get the second element of s; at 2, the third; and so on. This leads us to our definition

$$[X]$$

$$tail : seq[X] \rightarrowtail\!\!\!\!\!\!\,\,+ \; seq[X]$$

$$\forall s : seq[X] \bullet tail\ s = (\lambda\, n : \mathbb{N} \bullet n + 1) \mathbin{;} (\{1\} \lhd s)$$

which contains both the computation on the indices and the domain subtraction. The function *tail* is not defined for the empty sequence, so caution in its application is again advised.

One way of writing a *nonempty* sequence, to emphasise that it is indeed nonempty, is to insist that it has at least one element. The sequence

$$\langle x \rangle \frown s$$

is nonempty, no matter what s is. We use this trick in the following laws

$$\begin{aligned} head\ (\langle x \rangle \frown s) &= x \\ tail\ (\langle x \rangle \frown s) &= s \end{aligned}$$

For nonempty s, we can say that s is made up from its head and its tail

$$s = \langle head\ s \rangle \frown tail\ s \quad \text{for } s \neq \langle\rangle$$

Reversal

The reversal of a sequence is formed by taking its elements in reverse order

$$rev\ \langle t, u, v \rangle = \langle v, u, t \rangle$$

This is defined in what will be a familiar manner by now: we must transform the indices thus

$$(rev\ s)i = s(\#s - n + 1)$$

A general definition is given by

$$
\begin{array}{l}
[X] \\
\hline
rev : seq[X] \rightarrow seq[X] \\
\hline
\forall s : seq[X] \bullet \\
\quad rev\ s = (\lambda\, n : 1 \mathinner{\ldotp\ldotp} \#s \bullet \#s - n + 1) \fatsemi s
\end{array}
$$

Since our sequences are finite in length, *rev* is total and enjoys these pleasing properties

$$
\begin{aligned}
rev\ \langle\rangle &= \langle\rangle \\
rev\ \langle x\rangle &= \langle x\rangle \\
rev\ (s \frown t) &= (rev\ t) \frown (rev\ s) \\
rev\ (rev\ s) &= s
\end{aligned}
$$

Restriction

Often, we want to look at a sequence and "forget" about certain—for the moment irrelevant—elements. For example, if we are recording the behaviour of a variable in a system, we might use the following sets

$$
\begin{aligned}
R &\triangleq \{read.n \mid n \in \mathbb{N}\} \\
W &\triangleq \{write.n \mid n \in \mathbb{N}\} \\
Event &\triangleq R \cup W
\end{aligned}
$$

We might record the behaviour as

$$trace : seq[Event]$$

and a typical value of *trace* may be

$$trace = \langle write.7, write.3, read.3, write.20, read.20, read.20\rangle$$

Suppose that we want the subsequence of just the *write* events, in the correct order. This would be given by the trace *restricted* to the set W

$$trace \upharpoonright W = \langle write.7, write.3, write.20\rangle$$

As you can see, $trace \upharpoonright W$ contains just those elements of *trace* which are in W; moreover their relative order is preserved. Similarly,

$$trace \upharpoonright R = \langle read.3, read.20, read.20\rangle$$

Before we can give a definition of restriction, we need an auxiliary function that takes any function from $\mathbb{N}$ to X to a *sequence* s (a function from $\mathbb{N}$ to X with domain $1\,..\,\#s$). We call this function *squash*. Squashing the empty function gives the empty sequence, squashing a function with holes closes them up and makes sure that indexing starts at 1, whilst squashing a sequence leaves it unchanged

$$\begin{array}{l} squash\ \{\} = \langle\rangle \\ squash\ \{3 \mapsto a, 6 \mapsto a, 7 \mapsto b, 10 \mapsto a\} \\ \quad = \{1 \mapsto a, 2 \mapsto a, 3 \mapsto b, 4 \mapsto a\} \\ \quad = \langle a, a, b, a\rangle \\ squash\ \langle a, a, b, a\rangle = \langle a, a, b, a\rangle \end{array}$$

Define

$$\begin{array}{l} [X] \\ \hline squash : (\mathbb{N} \mathrel{+\!\!\!\twoheadrightarrow} X) \to seq[X] \\ \hline squash\ \{\} = \langle\rangle \\ \forall f : \mathbb{N} \mathrel{+\!\!\!\twoheadrightarrow} X; i : \mathbb{N} \mid f \neq \{\} \wedge i = min\ (dom\ f) \bullet \\ \quad squash\ f = \langle f\ i\rangle \frown squash(\{i\} \mathbin{\lhd\!\!\!-} f) \end{array}$$

Thus, if

$$f = \{1 \mapsto t, 3 \mapsto v\}$$

then

$$\begin{array}{l} squash\ f \\ \quad = squash\ \{1 \mapsto t, 3 \mapsto v\} \\ \quad = \langle t\rangle \frown squash\ (\{1\} \mathbin{\lhd\!\!\!-} \{1 \mapsto t, 3 \mapsto v\}) \\ \quad = \langle t\rangle \frown squash\ \{3 \mapsto v\} \\ \quad = \langle t\rangle \frown (\langle v\rangle \frown squash\ (\{3\} \mathbin{\lhd\!\!\!-} \{3 \mapsto v\})) \\ \quad = \langle t\rangle \frown (\langle v\rangle \frown squash\ \{\}) \\ \quad = \langle t\rangle \frown (\langle v\rangle \frown \langle\rangle) \\ \quad = \langle t\rangle \frown \langle v\rangle \\ \quad = \langle t, v\rangle \end{array}$$

A property of *squash* is that if the function to be squashed has no holes in its domain, then squashing it merely shifts its indices

$$dom\ f = l + 1 \ldots h \Rightarrow squash\ f = (\lambda\, n : \mathbb{N} \bullet n + l) \mathbin{;} f$$

Now, back to the definition of restriction, which is so called because it is rather like range restriction, appropriately squashed, of course

$$\begin{array}{|l}
\hline
[X] \\
_ \upharpoonright _ : seq[X] \times \mathbb{P}\, X \rightarrow seq[X] \\
\hline
\forall s : seq[X];\ A : \mathbb{P}\, X \bullet \\
\quad s \upharpoonright A = squash\ (s \rhd A) \\
\hline
\end{array}$$

So, if we take our backing store queue, and ignore all entries except v and t, we get

$$\begin{array}{rcl}
bsq \upharpoonright \{v, t\} & = & \langle t, u, v \rangle \upharpoonright \{v, t\} \\
& = & \{1 \mapsto t, 2 \mapsto u, 3 \mapsto v\} \upharpoonright \{v, t\} \\
& = & squash\ (\{1 \mapsto t, 2 \mapsto u, 3 \mapsto v\} \rhd \{v, t\}) \\
& = & squash\ \{1 \mapsto t, 3 \mapsto v\} \\
& = & \langle t, v \rangle
\end{array}$$

If we fix the set in question to, say, A, then restriction to A is strict and distributive in the sense described above

$$\begin{array}{rcl}
\langle\rangle \upharpoonright A & = & \langle\rangle \\
(s \frown t) \upharpoonright A & = & (s \upharpoonright A) \frown (t \upharpoonright A)
\end{array}$$

and its effect on the singleton sequence is determined by membership of A

$$\langle x \rangle \upharpoonright A = \begin{cases} \langle x \rangle & \text{if } x \in A \\ \langle\rangle & \text{if } x \notin A \end{cases}$$

The point about describing a function as both strict and distributive is that its effect on singleton sequences *uniquely* characterises it

$$\begin{array}{rcl}
bsq \upharpoonright \{v, t\} & = & \langle t, u, v \rangle \upharpoonright \{v, t\} \\
& = & (\langle t \rangle \frown \langle u \rangle \frown \langle v \rangle) \upharpoonright \{v, t\} \\
& = & (\langle t \rangle \upharpoonright \{v, t\}) \frown (\langle u \rangle \upharpoonright \{v, t\}) \frown (\langle v \rangle \upharpoonright \{v, t\}) \\
& = & \langle t \rangle \frown \langle\rangle \frown \langle v \rangle \\
& = & \langle t, v \rangle
\end{array}$$

Restriction to a set containing every member of a particular sequence won't change it

$$A \supseteq ran\ s \Rightarrow s \upharpoonright A = s$$

Restriction to a set containing no members of s reduces to the empty sequence

$$(ran\ s) \cap A = \{\} \Rightarrow s \upharpoonright A = \langle\rangle$$

an obvious consequence of which is that

$$s \upharpoonright \{\} = \langle\rangle$$

The only elements that can survive two ordeals by fire are those that can survive both

$$(s \upharpoonright A) \upharpoonright B = s \upharpoonright (A \cap B)$$

and finally, reversal and restriction may be performed in either order

$$(rev\ s) \upharpoonright A = rev\ (s \upharpoonright A)$$

Distributed Catenation

Just like the generalisation of union and intersection, we can imagine a generalisation of catenation. As we shall see in Section 8.7, the idea of sequences of sequences of objects is a useful one, and should be no stranger to us than the ideas of sets of sets or functions of functions. One useful operator on sequences of sequences is *distributed catenation*, or as it is often called—for obvious reasons—*flattening*.

Example 8.1 *When a sequence of sequences is flattened, the result consists of the constituent sequences catenated in order*

$$\begin{array}{l} ^\frown/\langle\langle a,b,c\rangle,\langle d,e\rangle\rangle = \langle a,b,c,d,e\rangle \\ ^\frown/\langle\langle\rangle,\langle\rangle,\langle\rangle,\langle\rangle\rangle = \langle\rangle \\ ^\frown/\langle\langle a\rangle,\langle a\rangle,\langle a\rangle\rangle = \langle a,a,a\rangle \\ ^\frown/\langle\langle\langle a,b,c\rangle,\langle d,e\rangle\rangle,\langle\langle f,g,h\rangle,\langle i,j,k\rangle\rangle\rangle = \\ \qquad \langle\langle a,b,c\rangle,\langle d,e\rangle,\langle f,g,h\rangle,\langle i,j,k\rangle\rangle \end{array}$$

Its definition is straightforward, since it is strict and distributive

$$
\begin{array}{l}
[X] \\
\frown/ : seq[seq[X]] \rightarrow seq[X] \\
\hline
\frown/\langle\rangle = \langle\rangle \\
\forall s : seq[X] \bullet \frown/\langle s \rangle = s \\
\forall ss_1, ss_2 : seq[seq[X]] \bullet \\
\qquad \frown/(ss_1 \frown ss_2) = (\frown/ss_1) \frown (\frown/ss_2)
\end{array}
$$

8.3 Some Special Sequences

Sequences have nice properties, and this probably accounts for their popularity. Often, however, we want particular sequences with special properties. For example, we might require a sequence which has no repeated elements, as in the backing store queuing example. In fact, this is just a restriction on the kind of function that we are using to model the sequence. In this case we want an *injection.*

$$
\begin{array}{l}
[X] \\
injseq : \mathbb{P}(seq[X]) \\
\hline
injseq = seq[X] \cap (\mathbb{N} \rightarrowtail\!\!\!\!\!\!+\; X)
\end{array}
$$

An injective sequence obviously has the property that we wanted

$$
\begin{array}{l}
\vdash \forall is : injseq[X] \bullet \\
\qquad \forall i, j : dom\ is \mid is\ i = is\ j \bullet i = j
\end{array}
$$

An example of an injective sequence is a *permutation*

$$
\begin{array}{l}
[X] \\
perm : \mathbb{F}\, X \rightarrow injseq[X] \\
\hline
\forall s : \mathbb{F}\, X \bullet ran\ (perm\ s) = s
\end{array}
$$

8.4 Partitions

An indexed set of sets S is said to *partition* a set T if the sets in S are *pairwise disjoint*, and between them contain all the elements of T. A useful example of an indexed set is a sequence—indexed by an initial segment of the natural numbers. Our definition formalised is

$$
\begin{array}{|l}
\_partition\_ : seq[\mathbb{P}\, X] \leftrightarrow \mathbb{P}\, X \\
\hline
\forall S : seq[\mathbb{P}\, X];\ T : \mathbb{P}\, X \bullet \\
\quad S\ partition\ T \Leftrightarrow \\
\quad\quad \forall i, j : dom\ S \mid i \neq j \bullet \\
\quad\quad\quad (S\ j) \cap (S\ j) = \{\} \\
\quad\quad\quad \bigcup\{i : dom\ S \bullet S\ i\} = T
\end{array}
$$

Example 8.2 *The following are example of partitions*

1. $\langle\{1\},\{2\},\{3\},\{4\}\rangle\ partition\ \{1,2,3,4\}$
2. $\langle\{1,2,3,4\}\rangle\ partition\ \{1,2,3,4\}$
3. $\langle\{1,2\},\{3,4\}\rangle\ partition\ \{1,2,3,4\}$
4. $\langle\{\},\{\},\{\},\{1,2,3,4\}\rangle\ partition\ \{1,2,3,4\}$

A simple law satisfied by our definition is

$$\langle A, B\rangle\ partition\ C \Leftrightarrow A \cap B = \{\} \wedge A \cup B = C$$

A more adventurous law is proved in the following theorem

Theorem 8.1 *Given a partition*

$$\langle S, T, U\rangle$$

of a set V, and a set $W \subseteq S$, we can form another partition of V by rearranging some of the elements of the partition thus

$$\langle S \setminus W, T, U \cup W\rangle$$

Proof

$$
\begin{array}{l}
\langle S, T, U\rangle\ partition\ V \Leftrightarrow \\
\quad S \cap T = \{\} \wedge S \cap U = \{\} \wedge T \cap U = \{\} \wedge \\
\quad S \cup T \cup U = V
\end{array}
$$

Consider the three sets

$$S \setminus W, T, U \cup W$$

They are pairwise disjoint

$$(S \setminus W) \cap T = (S \cap T) \setminus W = \{\} \setminus W = \{\}$$

$$\begin{aligned}
&(S \setminus W) \cap (U \cup W) \\
&\quad = (S \cap (U \cup W)) \setminus W \\
&\quad = ((S \cap U) \cup (S \cap W)) \setminus W \\
&\quad = (\{\} \cup (S \cap W)) \setminus W \\
&\quad = (S \cap W) \setminus W \\
&\quad = W \setminus W \\
&\quad = \{\}
\end{aligned}$$

$$\begin{aligned}
&T \cap (U \cup W) \\
&\quad = (T \cap U) \cup (T \cap W) \\
&\quad = \{\} \cup (T \cap W) \\
&\quad = T \cap W \\
&\quad = W \cap T \\
&\quad \subseteq S \cap T \\
&\quad = \{\}
\end{aligned}$$

and their union is the whole of V

$$\begin{aligned}
&(S \setminus W) \cup T \cup (U \cup W) \\
&\quad = (S \setminus W) \cup T \cup U \cup W \\
&\quad = (S \setminus W) \cup W \cup T \cup U \\
&\quad = S \cup T \cup U \\
&\quad = V
\end{aligned}$$

QED

8.5 Ordering

Suppose that s and t are both sequences of X, and furthermore, that

$$s \subseteq t$$

If $\#s = n$, then we can see that s is just the first n elements of t

$$s = (1 \mathinner{..} n) \lhd t = (dom\ s) \lhd t$$

The proof of this is really very simple; first, recall that

$$s \subseteq t \quad \Leftrightarrow \quad \forall i : \mathbb{N}; x : X \mid i \in dom\ s \bullet s\ i = t\ i$$

Now, expanding $(dom\ s) \lhd t$, we get

$$\begin{aligned}
&(dom\ s) \lhd t \\
&\quad = \{i : \mathbb{N}; x : X \mid i \in dom\ s \wedge t\ i = x \bullet i \mapsto x\} \\
&\quad = \{i : \mathbb{N}; x : X \mid i \in dom\ s \wedge s\ i = x \bullet i \mapsto x\} \\
&\quad = s
\end{aligned}$$

Theorem 8.2 *If a sequence s is a subset of another sequence t, then we can always find an extension to s to make up the whole of t*

$$\vdash s, t : seq[X] \mid s \subseteq t \Rightarrow \exists u : seq[X] \bullet s \frown u = t$$

Proof: We can express t as the union of two relations

$$t = ((dom\ s) \lhd t) \cup ((dom\ s) ⩤ t)$$

But since $s \subseteq t$, the first of these two relations is simply s, so

$$t = s \cup ((dom\ s) ⩤ t)$$

The domain of $(dom\ s) ⩤ t$ is

$$\begin{aligned}
&(dom\ t) \setminus (dom\ s) \\
&\quad = (1 \mathinner{..} \#t) \setminus (1 \mathinner{..} \#s) \\
&\quad = (\#s + 1) \mathinner{..} \#t
\end{aligned}$$

So,

$$((dom\ s) \lhd t) = Id_{(\#s+1)..\#t} \mathbin{;} ((dom\ s) \lhd t)$$

Now, we can rewrite this identity function as follows

$$\begin{aligned} & Id_{(\#s+1)..\#t} \\ & \quad = \lambda\, n : \mathbb{N} \mid \#s < n \leq \#t \bullet n \\ & \quad = \lambda\, n : \mathbb{N} \mid \#s < n \leq \#t \bullet (n - \#s) + \#s \\ & \quad = (\lambda\, n : \mathbb{N} \mid \#s < n \leq \#t \bullet n - \#s) \mathbin{;} (\lambda\, n : \mathbb{N} \bullet n + \#s) \end{aligned}$$

Substituting this expansion into our expression for $(dom\ s) \lhd t$ we obtain

$$\begin{aligned} & (dom\ s) \lhd t \\ & \quad = (\lambda\, n : \mathbb{N} \mid \#s < n \leq \#t \bullet n - \#s) \mathbin{;} \\ & \qquad\quad (\lambda\, n : \mathbb{N} \bullet n + \#s) \mathbin{;} ((dom\ s) \lhd t) \\ & \quad = (\lambda\, n : \mathbb{N} \mid \#s < n \leq \#t \bullet n - \#s) \mathbin{;} squash((dom\ s) \lhd t) \end{aligned}$$

We now have a useful expression for the second of the two relations that united constitute t

$$t = s \cup ((\lambda\, n : \mathbb{N} \mid \#s < n \leq \#t \bullet n - \#s) \mathbin{;} squash((dom\ s) \lhd t))$$

But from the definition of catenation we can rewrite this as

$$t = s \frown squash\ (dom\ s \lhd t)$$

QED.

When s and t are in this relation, we say that s is *an initial subsequence* or *a prefix* of t[3]. Examples of this are

$$\begin{aligned} \langle t, u \rangle & \subseteq \langle t, u, v \rangle \\ \langle write.7 \rangle & \subseteq \langle write.7, write.3, read.3 \rangle \end{aligned}$$

This relation on sequences is known as a *partial ordering*, because it enjoys the following properties, inherited from the definition of a sequence as a function. There is a *least element*

$$\langle\rangle \subseteq s$$

[3] This prefix order is sometimes written as $s \leq t$.

The prefix ordering is clearly *reflexive*, that is

$$s \subseteq s$$

It is *antisymmetric*; if two sequences are prefixes of each other, then they are one and the same

$$s \subseteq t \wedge t \subseteq s \Rightarrow s = t$$

Finally, the ordering is *transitive*

$$s \subseteq t \wedge t \subseteq u \Rightarrow s \subseteq u$$

Strictly speaking, it is the set of sequences over X and the prefix ordering considered as a pair $(seq[X], \subseteq)$ which forms the partially ordered set, although common practice has it that $seq[X]$ is a partially ordered set, $\subseteq$ being implicitly understood. We use the word *partial* to describe the ordering because we have chosen not to order certain elements. For instance,

$$\langle t, u, v, w \rangle \not\subseteq \langle t, u, v, x \rangle \quad \text{and} \quad \langle t, u, v, x \rangle \not\subseteq \langle t, u, v, w \rangle$$

if $x \neq w$. Therefore, the prefix ordering is not *total*.

8.6 Proof by Induction

The most basic property of the natural numbers is that of *mathematical induction*. Suppose that we want to prove as a theorem

$$\vdash \forall n : \mathbb{N} \bullet P(n)$$

There are three strategies open to us

1. Show that $P(n)$ holds for every n in $\mathbb{N}$ by considering each in turn. This is not in general a very useful suggestion.

2. Show that $\neg \exists n : \mathbb{N} \bullet \neg P(n)$, by assuming there to be a value in $\mathbb{N}$ such that $\neg P(n)$, and deriving a contradiction.

3. Use an arbitrary value to obtain $P(a)$, and then introduce universal quantification.

A proof by induction that $P(n)$ is true for all n consists of proving the *basis* of the induction $P(0)$, and the *inductive step* that

$$P(n) \Rightarrow P(n+1)$$

If we have proved these two things, then we can easily establish the truth of $P(k)$, for some k, by starting from $P(0)$, and applying the inductive step as often as necessary to reach $P(k)$.

Example 8.3 *Induction can be used to prove many of the properties of the natural numbers, such as the following*

$$\sum_{i=0}^{n} i = 0 + 1 + 2 + 3 + \cdots = \tfrac{1}{2} \times n \times (n+1)$$

Proof: The basis of the proof is true by definition

$$\sum_{i=0}^{0} i = 0$$

For the inductive step, we may assume that

$$\sum_{i=0}^{k} i = \tfrac{1}{2} \times k \times (k+1)$$

Expanding the formula for $k+1$ in terms of k gives

$$\begin{aligned}\sum_{i=0}^{k+1} i &= \sum_{i=0}^{k} i + (k+1)\\ &= \tfrac{1}{2} \times k \times (k+1) + (k+1)\\ &= (k+1) \times (\tfrac{1}{2} \times k + 1)\\ &= (k+1) \times \tfrac{1}{2} \times (k+2)\\ &= \tfrac{1}{2} \times (k+1) \times (k+2)\end{aligned}$$

QED

Exercise 8.2 *Prove by induction that*

$$\sum_{i=0}^{n} i^2 = \tfrac{1}{6} \times n \times (n+1) \times (2 \times n + 1)$$

Exercise 8.3 *Prove by induction that*

$$\sum_{i=0}^{n} i^3 = (\sum_{i=0}^{n} i)^2$$

Exercise 8.4 *Given the definitions*

$$a^1 = a$$
$$a^{n+1} = a^n \times a$$

prove by induction that

$$a^{m+n} = a^m \times a^n$$
$$(a^m)^n = a^{m \times n}$$

Exercise 8.5 *Given the definitions*

$$a + (b + c) = (a + b) + c$$
$$a + b = b + a$$
$$1 \times b = b$$
$$(a + 1) \times b = a \times b + b$$

prove the following correct

$$a \times (b + c) = a \times b + a \times c$$
$$a \times b = b \times a$$

Structural Induction

We can also use induction over certain structures, such as finite sets and sequences, in much the same manner as we have for the natural numbers. There are several ways to formulate the induction principle for sequences, and we shall consider the simplest. Suppose that we want to prove that $P(s)$ is true for all sequences s. The basis is established by showing that $P(\langle\rangle)$. The inductive step requires that we prove that $P(s) \Rightarrow P(\langle x \rangle \frown s)$.

Example 8.4 *We shall prove by structural induction the theorem*

$$rev\ (rev\ s) = s$$

The proof follows from the following three laws about reversal (that it is strict, what it does to the unit sequence, and how it distributes through catenation)

$$
\begin{aligned}
&rev\ \langle\rangle = \langle\rangle \\
&rev\ \langle x\rangle = \langle x\rangle \\
&rev\ (s \frown t) = (rev\ t) \frown (rev\ s)
\end{aligned}
$$

Proof: First we consider the basis

$$rev\ (rev\ \langle\rangle) = rev\ \langle\rangle = \langle\rangle$$

And now for the inductive step, assuming that $rev\ (rev\ s) = s$

$$
\begin{aligned}
&rev\ (rev\ (\langle x\rangle \frown s)) \\
&\quad = rev\ ((rev\ s) \frown (rev\ \langle x\rangle)) \\
&\quad = rev\ ((rev\ s) \frown \langle x\rangle) \\
&\quad = (rev\ \langle x\rangle) \frown rev\ (rev\ s) \\
&\quad = \langle x\rangle \frown rev\ (rev\ s) \\
&\quad = \langle x\rangle \frown s
\end{aligned}
$$

QED

Exercise 8.6 *Example 8.4 used the following law to prove that rev is an* involution

$$rev\ (s \frown t) = (rev\ t) \frown (rev\ s)$$

Prove that it really is a law.

Exercise 8.7 *Prove first by induction, and then by algebraic manipulations that* $\#(s \frown t) = \#s + \#t$.

Exercise 8.8 *Using structural induction, prove that the length of the reversal of a sequence is just the same as the length of the sequence itself.*

8.7 Sequences of Sequences of...

In this section we present a mathematical theory about objects that we call *Tables*. Developing this theory makes the example in Section 8.7 much easier to describe. This is often the case with specification work: time spent understanding what you are talking about reduces the amount of time you need to spend talking!

A Theory of Tables

A sequence is a one dimensional object—the dimension being the order of the elements. We can easily imagine higher dimensions by taking sequences over sequences. For example, we can describe a two dimensional table of numbers as a sequence of rows, each row being a sequence of numbers. As we have in mind a rectangular object, we shall impose the condition that each row is of equal length

$$[X]$$

$$Table[X] : \mathbb{P}(seq[seq[X]])$$

$$Table = \{t : seq[seq[X]] \mid \forall s_1, s_2 : ran\ t \bullet \#s_1 = \#s_2\}$$

Now we choose the ith row of a table t (for $i \in dom\ t$) to be simply the ith sequence in t, $t\ i$, and the jth element of the ith row to be $t\ i\ j$.

We shall define a couple of functions on tables to determine the row and column indices

$$[X]$$

$$rows, cols : Table[X] \to \mathbb{P}\,\mathbb{N}$$

$$\forall t : Table[X] \bullet$$
$$rows\ t = dom\ t$$
$$cols\ t = dom\ (\bigcup(ran\ t))$$

Since the rows are all of the same length

$$dom\ (\bigcup(ran\ t))$$

really does give us the column indices

$$\vdash \forall t : Table[X] \bullet \forall s : ran\ t \bullet dom\ s = cols\ t$$

which is a segment of the natural numbers, starting at 1

$$\vdash \forall t : Table[X] \bullet \exists n : \mathbb{N} \bullet cols\ t = 1 \mathinner{..} n$$

Furthermore, every pair of row and column indices identifies a table entry

$$\begin{array}{l} \vdash \forall t : Table[X] \bullet \\ \quad \forall i : rows\ t; j : cols\ t \bullet \\ \qquad i \in dom\ t \wedge j \in dom\ (t\ i) \end{array}$$

$$\begin{array}{l} \vdash \forall t : Table[X] \bullet \\ \quad \forall i : rows\ t; j : cols\ t \bullet \\ \qquad \exists n : \mathbb{N} \bullet n = t\ i\ j \end{array}$$

How can we select the jth *column*? Let's define a function to do this for us, which we call *ref*: it should take a table and a column index, and return a sequence of X—the column. What should this function do? It allows us to refer to an element by providing the indices in *reverse* order. So,

$$ref\ t\ j\ i = t\ i\ j$$

The definition of *ref* shows how we permute the indices

$$ref \mathrel{\widehat{=}} \lambda t : Table[X] \bullet (\lambda j : cols\ t \bullet (\lambda i : rows\ t \bullet t\ i\ j))$$

It is clear that

$$ref \in Table[X] \pfun \mathbb{N} \pfun \mathbb{N} \pfun X$$

but we can say more than this. Having specified a table t and a column j, the result is more than a function from the natural numbers to X, it is a *sequence over X*

$$\lambda i : rows\ t \bullet t\ i\ j$$

Theorem 8.3 *The jth column of a table t is a sequence over X*

$$\begin{array}{l} \forall t : Table[X] \bullet \\ \quad \forall j : cols\ t \bullet \\ \quad\quad (\lambda\, i : rows\ t \bullet t\ i\ j) \in seq[X] \end{array}$$

Proof: Clearly,

$$(\lambda\, i : rows\ t \bullet t\ i\ j) \in \mathbb{N} \twoheadrightarrow X$$

Consider

$$\begin{array}{rcl} dom\ (\lambda\, i : rows\ t \bullet t\ i\ j) & = & dom\ \{i : rows\ t \bullet i \mapsto t\ i\ j\} \\ & = & dom\ \{i : dom\ t \bullet i \mapsto t\ i\ j\} \\ & = & \{i : dom\ t \mid \exists\, n : X \bullet n = t\ i\ j\} \\ & = & \{i : dom\ t \mid true\} \\ & = & dom\ t \\ & = & 1 \mathrel{..} \#t \end{array}$$

Therefore

$$(\lambda\, i : rows\ t \bullet t\ i\ j) \in seq[X]$$

QED

Also, perhaps with a little more thought, we can see that

$$(\lambda\, j : cols\ t \bullet \lambda\, i : rows\ t \bullet t\ i\ j) \in seq[seq[X]]$$

but this should be so, since each column is a sequence, and a table is also a sequence of columns.

Theorem 8.4 *Every table can be represented as a sequence of columns, instead of as a sequence of rows*

$$\begin{array}{l} \forall t : Table[X] \bullet \\ \quad (\lambda\, j : cols\ t \bullet \lambda\, i : rows\ t \bullet t\ i\ j) \in seq[seq[X]] \end{array}$$

Proof: Clearly, from the previous theorem,

$$\begin{array}{ll} & \forall i : rows\ t; j : cols\ t \bullet i \in dom\ t \wedge j \in dom\ (t\ i) \\ \Leftrightarrow & \forall j : cols\ t \bullet \exists\, s : \mathbb{N} \twoheadrightarrow X \bullet s = \{i : rows\ t \bullet i \mapsto t\ i\ j\} \\ \Leftrightarrow & \forall j : cols\ t \bullet \exists\, s : \mathbb{N} \twoheadrightarrow X \bullet s = (\lambda\, i : rows\ t \bullet t\ i\ j) \\ \Leftrightarrow & \forall j : cols\ t \bullet \exists\, s : seq[X] \bullet s = (\lambda\, i : rows\ t \bullet t\ i\ j) \end{array}$$

Now consider

$$\begin{array}{rl} & dom\ (\lambda\, j : cols\ t \bullet \lambda\, i : rows\ t \bullet t\ i\ j) \\ = & dom\ \{j : cols\ t \bullet j \mapsto (\lambda\, i : rows\ t \bullet t\ i\ j)\} \\ = & \{j : cols\ t \mid \exists\, s : seq[X] \bullet s = (\lambda\, i : rows\ t \bullet t\ i\ j)\} \\ = & \{j : cols\ t \mid true\} \\ = & cols\ t \end{array}$$

but since $\exists\, n : \mathbb{N} \bullet cols\ t = 1 \mathrel{..} n$*, we have*

$$(\lambda\, j : cols\ t \bullet \lambda\, i : rows\ t \bullet t\ i\ j) \in seq[seq[X]]$$

QED

Finally, the following theorem holds good

Theorem 8.5 *A reflected table is still a table*

$$ref \in Table[X] \rightarrow Table[X]$$

Proof: An exercise for the reader.

Hence the choice of name for our function: *ref* is for *reflect*. It takes a table and reflects it about the diagonal which starts in the top left corner.

Exercise 8.9 *An obviously desirable property is that*

$$ref\ (ref\ t) = t$$

Prove this.

Magic Squares

A *magic square* is a table of numbers such as those in Figures 8.2, 8.3, and 8.4. A magic square of size n is characterised by being *square* (having equal numbers of rows and columns), by containing all the numbers between 1 and n^2 as its entries, and having all its columns, rows, and major diagonals add up to the same sum. In order to formalise this, we use the theory of tables and the following definitions. The *sum* over a sequence of numbers is simply the sum of all its entries

2	9	4
7	5	3
6	1	8

Figure 8.2: A 3×3 magic square

25	18	11	9	2
14	7	5	23	16
3	21	19	12	10
17	15	8	1	24
6	4	22	20	13

Figure 8.3: A 5×5 magic square

4	29	12	37	20	45	28
35	11	36	19	44	27	3
10	42	18	43	26	2	34
41	17	49	25	1	33	9
16	48	24	7	32	8	40
47	23	6	31	14	39	15
22	5	30	13	38	21	46

Figure 8.4: A 7×7 magic square

$$\begin{array}{|l} sum : seq[\mathbb{N}] \rightarrow \mathbb{N} \\ \hline sum\ \langle\rangle = 0 \\ sum\ \langle n \rangle = n \\ sum\ (s \frown t) = (sum\ s) + (sum\ t) \end{array}$$

Sq is the set of *square Tables* of numbers

$$\begin{array}{|l} Sq : \mathbb{P}\ Table[\mathbb{N}] \\ \hline Sq = \{sq : Table[X] \mid rows\ s = cols\ s\} \end{array}$$

The leading diagonal of a square is the one from top left to bottom right

$$ldiag \mathrel{\widehat{=}} \lambda\, sq : Sq \bullet \lambda\, i : dom\ sq \bullet sq\ i\ i$$

The other diagonal runs from the bottom left to the top right

$$diag \mathrel{\widehat{=}} \lambda\, sq : Sq \bullet \lambda\, i : dom\ sq \bullet sq\ (\#sq + 1 - i)\ i$$

Magic squares are drawn from the following

$$\begin{array}{|l} magicSq : \mathbb{N} \twoheadrightarrow \mathbb{P}\ Sq \\ \hline \forall\, n : \mathbb{N};\ sq : Sq \bullet \\ \quad sq \in magicSq\ n \Leftrightarrow \\ \qquad \frown/sq \in perm\ (1 \mathinner{..} n^2) \wedge \\ \qquad \exists\, m : \mathbb{N} \bullet \\ \qquad\quad \forall\, i : rows\ sq \bullet \\ \qquad\qquad sum\ (sq\ i) = m \wedge \\ \qquad\qquad sum\ (ref\ sq\ i) = m \wedge \\ \qquad\qquad sum\ ldiag\ sq = m \wedge \\ \qquad\qquad sum\ diag\ sq = m \end{array}$$

8.8 Summary

We have introduced the last of our "mathematical data types": the sequence. Sequences allow us to express a collection of ordered objects, possibly with repetition, and they have some nice properties and a rich set of operators. Sequences are particularly useful in describing software systems, and we have given some case studies where they play an important part. Just as with relations and functions, sequences are built on top of the set theory that we introduced in Chapter 5; all the descriptions that we have given in this chapter could equally well have been given using simple set theory. The point is that we have developed a theory of sequences to make our life easier. This is a lesson that we can generalise: whenever we use mathematics to specify something, we should first develop an application-oriented theory to make things easier for ourselves.

Chapter 9

Case Study: A Telephone Exchange

It might be said that many books on formal methods should really be found in an intellectual nursery: they use the might of mathematics to treat trivial problems from first principles. In this book we have tried to give examples of the use of our mathematics applied to genuine problems in software development. These included a configuration manager for a programming support environment, a storage allocator for an operating system, and a queuing mechanism for a backing store interface.

In this chapter we shall give a more sustained case study, we shall apply our mathematical tools to the specification of a real-life object with which we are familiar: a telephone exchange[1]. Because the case study is quite a large one, it must be read carefully and worked through before it can be fully understood. Real-life systems can be quite complicated, and we have tried to subordinate this complexity with our mathematical descriptions. The exercises are almost all proofs of lemmas and theorems about the system, and completing these exercises will give more insight into the system and its description. It is quite probable that the reader will be able to propose an

[1]This description is based on that in Woodcock, J.C.P., "The Formal Specification of a Simple PABX Telephone Exchange", Internal Report, GEC Research Laboratories.

alternative model for the telephone exchange, perhaps because the reader is familiar with a different system and wants to incorporate its properties, or perhaps because the reader can think of a model which expresses the existing properties more conveniently. This is encouraged.

9.1 Subscribers and Telephone Calls

Consider a small telephone system, such as might be installed internally in an office building. We shall model the behaviour of the *telephone exchange* at the heart of this system. We start by describing the *state* of the exchange, that is, the collection of objects upon which the exchange operates. The state contains all the information which characterises an exchange at any instant of time. Later we shall describe how various *operations* which may be invoked by users of the telephone system affect the state of the exchange.

In essence, telephone exchanges are all about *subscribers*: the users of the telephone system. When one subscriber telephones another the interface requires that the second subscriber's "network address" is known by the first. In other words, in order to telephone someone you need to know their *telephone number*. This number is input to the system digit by digit. This is true regardless of the subscriber's equipment, no matter whether the user dials the number, nor whether the number is stored and retrieved at the stroke of a key. Therefore we shall need to know something about the structure of the set of all subscribers: a subscriber is a sequence of digits. Let *Subs* be the set of all subscribers

$$Subs : \mathsf{P}(seq[Digit])$$

where *Digit* is defined to be

$$Digit ::= 0 \mid 1 \mid 2 \mid 3 \mid 4 \mid 5 \mid 6 \mid 7 \mid 8 \mid 9$$

It would be over-specifying if we attempted to say exactly which digit sequences were subscribers; it is enough to know that certain digit sequences correspond to *unobtainable* numbers, and some are neither subscribers, nor are they unobtainable.

When subscribers initiate a call the first thing that they do is to lift their handset—or do something else to attract the system's attention: in telephony jargon this is known as *seizing a line*. Next they have to tell the system who they are trying to call: this is the process of *dialling*. The number they are trying to reach might be *unobtainable*, but if not, the system will be *connecting* the call. Depending on the called subscriber, the state of the call will become either *ringing* or *engaged*. Once a called subscriber answers a ringing call, *speech* ensues—we suppose! The telephone system that we are modelling has few frills, but there is a requirement that only the subscriber who initiated a call can terminate it. If the called subscriber tries to terminate a call, the call becomes *suspended*.

The status of a call can be any of the following

$$\begin{aligned}&\mathit{Status} ::= \\ &\quad \mathit{seize} \mid \mathit{dialling} \mid \mathit{unobtainable} \mid \mathit{connecting} \mid \\ &\quad \mathit{ringing} \mid \mathit{speech} \mid \mathit{engaged} \mid \mathit{suspended}\end{aligned}$$

We can say a little more about subscribers and the status of a telephone call. If a subscriber is in the process of dialling another subscriber's number, only certain sequences of digits so far dialled are valid—that is, can lead to a subscriber's number. A sequence of digits is valid only if it is a prefix of some subscriber

$$\begin{array}{|l}\mathit{Valid} : \mathbb{P}(\mathit{seq}[\mathit{Digit}]) \\ \hline \mathit{Valid} = \{n : \mathit{seq}[\mathit{Digit}] \mid (\exists\, s : \mathit{Subs} \bullet n \subseteq s) \bullet n\}\end{array}$$

Valid is the *prefix closure* of the set of subscribers *Subs*.

Theorem 9.1 *Obviously, every subscriber has a valid number*

$$\mathit{Subs} \subseteq \mathit{Valid}$$

We say that a call is *Connected* if it is one of the following: *ringing*, *speech*, or *suspended*

$$\mathit{Connected} \mathrel{\widehat=} \{\mathit{ringing}, \mathit{speech}, \mathit{suspended}\}$$

and it is *Established* if it is *Connected*, *connecting*, or *engaged*

$$\mathit{Established} \mathrel{\widehat=} \mathit{Connected} \cup \{\mathit{connecting}, \mathit{engaged}\}$$

Now we are in a position to say something about the relation between a sequence of digits dialled so far in a call and the possible status of the call. The state *seize* is one where no digits have been dialled; *unobtainable* is the state where the number dialled is invalid[2]; the *Established* states are those where the number dialled is a subscriber's number. The relation *SubRec* summarises this

$$
\begin{array}{|l}
SubRec : Status \leftrightarrow seq[Digit] \\
\hline
\forall s : Status; n : seq[Digit] \bullet \\
\quad (s, n) \in SubRec \Leftrightarrow \\
\quad\quad s = seize \Leftrightarrow n = \langle\rangle \\
\quad\quad s = unobtainable \Leftrightarrow n \notin Valid \\
\quad\quad s \in Established \Leftrightarrow n \in Subs
\end{array}
$$

We shall find a use for the two projection functions on *SubRec*: *st* and *num* which give the status and number of a *SubRec* pair

$$
\begin{array}{|l}
st : SubRec \rightarrow Status \\
num : SubRec \rightarrow seq[Digit] \\
\hline
\forall s : Status; n : seq[Digit] \mid (s, n) \in SubRec \bullet \\
\quad st(s, n) = s \wedge num(s, n) = n
\end{array}
$$

An obvious property of the two projection functions is that

$$\forall sr : SubRec \bullet sr = (st\ sr, num\ sr)$$

Theorem 9.2 *If a subscriber is in the process of dialling, then the number so far dialled is valid, but is not a subscriber's number*

$$
\begin{array}{l}
\forall n : seq[Digit] \mid (dialling, n) \in SubRec \bullet \\
\quad n \in (Valid \setminus Subs) \setminus \{\langle\rangle\}
\end{array}
$$

Proof

$$
\begin{array}{l}
(dialling, n) \in SubRec \\
\quad \Leftrightarrow \ (dialling = seize \Leftrightarrow n = \langle\rangle) \wedge \\
\quad\quad\ \ (dialling = unobtainable \Leftrightarrow n \notin Valid) \wedge \\
\quad\quad\ \ (dialling \in Established \Leftrightarrow n \in Subs)
\end{array}
$$

[2]That is, is not in *Valid*!

$$\Leftrightarrow n \neq \langle\rangle \wedge n \in Valid \wedge n \notin Subs$$
$$\Leftrightarrow n \in Valid \setminus Subs \wedge n \neq \langle\rangle$$
$$\Leftrightarrow n \in (Valid \setminus Subs) \setminus \{\langle\rangle\}$$

QED

We could have included another conjunct in the definition of *SubRec*, namely

$$s = dialling \Leftrightarrow n \in (Valid \setminus Subs) \setminus \{\langle\rangle\}$$

This would have involved proving the consistency of the definition instead of proving Theorem 9.2.

9.2 The State of an Exchange

In our telephone exchange, subscribers may be thought of as being *Free*, *Unavailable*, or busy. If a subscriber is either free or unavailable, then, for the purposes of this level of specification, we do not need to know any other information about them. We can introduce the sordid details of accounting and so on at a later stage. We can therefore think of the free and unavailable subscribers as sets in our state, since all we are interested in is their identity. We shall represent their type as the power set of the set of all subscribers

$$Free, Unavailable : \mathbb{P}\ Subs$$

For busy subscribers—those in the middle of a telephone call—we require a little more information, namely some data recording the number that has been dialled, together with the status of the call. Of course, the two pieces of data must be in the relation that we called *SubRec*. The function *call* will be used to keep track of calls in progress

$$call : Subs \nrightarrow SubRec$$

Since *SubRec* is a relation between *Status* and *seq*[*Digit*], its type is really

$$\mathbb{P}(Status \times seq[Digit])$$

An element of *SubRec* is therefore of type

$$Status \times seq[Digit]$$

So, for $s \in Subs$

$$(call\ s) \in Status \times seq[Digit]$$

and is thus a pair. Note that our definition means that each subscriber can have at most one call in progress, since *call* is functional. The function *call* has in its domain all those initiators of calls. Another set of interest is that containing all those initiators who have succeeded in connecting. If we compose *call* with the projector *st*

$$call \fatsemi st$$

we get a map from initiators to the states of their calls. We can further constrain this to tell us about only those initiators who have succeeded in connecting their calls

$$(call \fatsemi st) \rhd Connected$$

This maps connected initiators to the states of their calls. We are interested particularly in the domain of this function

$$Callers : \mathbb{P}\ Subs$$

$$Callers = dom\ ((call \fatsemi st) \rhd Connected)$$

Of course we should just make a small check to make sure that all callers really are subscribers.

Theorem 9.3 *All callers are subscribers*

$$dom\ ((call \fatsemi st) \rhd Connected) \subseteq Subs$$

Proof

$$\begin{aligned} & dom\ ((call \fatsemi st) \rhd Connected) \\ & \quad \subseteq dom\ (call \fatsemi st) \\ & \quad \subseteq dom\ call \\ & \quad \subseteq Subs \end{aligned}$$

QED

Theorem 9.3 tells us that we never need to check to make sure that *Callers* is a subset of *Subs*: so long as *call* is as we have described, *Callers* will invariantly be a subset of *Subs*.

An important part of the state of a telephone exchange is the relation between connected subscribers. Here we have some requirements forced upon us: no subscriber may be successfully connected to more than one other subscriber. This goes for both the initiator and the recipient[3], so our relation *connected* must be an injective function

$$connected : Subs \rightarrowtail\!\!\!\!\!\!\!\!\;\shortmid\;\; Subs$$

To define what *connected* actually is, consider the mapping

$$call \fatsemi num$$

It relates initiators to the numbers they have dialled so far. We want only successful initiators in the domain of our mapping

$$connected = Callers \lhd (call \fatsemi num)$$

We have done some fancy footwork to construct *connected*; how do we know that it is a relation between subscribers and not just digit sequences?

Theorem 9.4 *Connections may only be made between subscribers*

$$\begin{array}{l} dom\ (Callers \lhd (call \fatsemi num)) \subseteq Subs \\ ran\ (Callers \lhd (call \fatsemi num)) \subseteq Subs \end{array}$$

Proof

$$\begin{array}{l} dom\ (Callers \lhd (call \fatsemi num)) \\ \quad = Callers \cap (dom\ (call \fatsemi num)) \\ \quad \subseteq Callers \\ \quad \subseteq Subs \end{array}$$

---

[3]Thus ruling out fancy telephone services such as *teleconferencing* and crossed lines!

Consider

$$\begin{aligned}
&ran\ ((call \mathbin{;} st) \rhd Connected)\\
&\quad = (ran\ (call \mathbin{;} st)) \cap Connected\\
&\quad \subseteq Connected\\
&\quad \subseteq Established
\end{aligned}$$

So every Connected call is Established. Now consider what the called number could be

$$\begin{aligned}
&\forall s : dom((call \mathbin{;} st) \rhd Connected) \bullet\\
&\qquad ((call \mathbin{;} st)\ s, (call \mathbin{;} num)\ s) \in SubRec\\
&\Leftrightarrow \forall s : dom((call \mathbin{;} st) \rhd Connected) \bullet\\
&\qquad (call \mathbin{;} num)\ s \in Subs\\
&\Leftrightarrow ran\ (call \mathbin{;} num) \subseteq Subs
\end{aligned}$$

QED

Theorem 9.4 tells us that *connected* is always a relation on *Subs*. Of course, there is nothing inherent about the relation that makes it *necessarily* functional, let alone injective; that fact must be checked whenever *connected* is altered.

Besides the subscribers who have initiated calls, there is another set of users who are busy: those who are recipients of telephone calls initiated by another subscriber. The set of recipients is nothing more than the range of *connected*.

The set of subscribers is partitioned by the free, unavailable and busy sets

$$\langle Free, Unavailable, (dom\ call) \cup (ran\ connected)\rangle\ partition\ Subs$$

The complete telephone exchange state is

$$\begin{array}{|l}
\textit{Exchange} \\
\hline
Free, Unavailable, Callers : \mathbb{P}\, Subs \\
call : Subs \pfun SubRec \\
connected : Subs \pinj Subs \\
\hline
\langle Free, Unavailable, (\mathrm{dom}\ call) \cup (\mathrm{ran}\ connected)\rangle \\
\quad \mathrm{partition}\ Subs \\
Callers = \mathrm{dom}\ ((call \comp st) \rres Connected) \\
connected = Callers \dres (call \comp num) \\
\hline
\end{array}$$

Some Consequences of the Definition of the State

We can bolster our confidence in our description by proving a few consequences of our definitions.

Theorem 9.5 *A free subscriber is neither making nor receiving a telephone call*

$$Free \cap (\mathrm{dom}\ call) = \{\} \land Free \cap (\mathrm{ran}\ connected) = \{\}$$

Proof

$$\begin{aligned}
&\langle Free, Unavailable, (\mathrm{dom}\ call) \cup (\mathrm{ran}\ connected)\rangle\ \mathrm{partition}\ Subs \\
&\qquad \Rightarrow Free \cap ((\mathrm{dom}\ call) \cup (\mathrm{ran}\ connected)) = \{\} \\
&\qquad \Rightarrow (Free \cap (\mathrm{dom}\ call)) \cup (Free \cap (\mathrm{ran}\ connected)) = \{\} \\
&\qquad \Rightarrow Free \cap (\mathrm{dom}\ call) = \{\} \land Free \cap (\mathrm{ran}\ connected) = \{\}
\end{aligned}$$

QED

Theorem 9.6 *A free subscriber isn't a connected caller*

$$Free \cap Callers = \{\}$$

Proof

Consider call; from the properties of $\comp$ *and* $\rres$ *we have*

$$\mathrm{dom}\ call \supseteq \mathrm{dom}\ ((call \comp st) \rres Connected) = Callers$$

Therefore

$$Free \cap Callers \subseteq Free \cap (\mathrm{dom}\ call) = \{\}$$

QED

9.3 The Initial Exchange State

Initially all subscribers are free and available, and there are no calls in progress

$$
\begin{array}{|l}
\hline
\textit{InitExchange} \\
\textit{Exchange}' \\
\hline
\textit{Free}' = \textit{Subs} \\
\textit{Unavailable}' = \{\} \\
\textit{call}' = \{\} \\
\hline
\end{array}
$$

Our exchange wouldn't be much use if it had no initial state.

Theorem 9.7 *An initial state exists for the telephone exchange.*

Consider the schema that we shall call InitExist, which is obtained by existentially quantifying the state variables in InitExchange

$$
\begin{array}{|l}
\hline
\textit{InitExist} \\
\exists\, \textit{Free}', \textit{Unavailable}', \textit{Callers}' : \mathbb{P}\, \textit{Subs}; \\
\quad \textit{call}' : \textit{Subs} \nrightarrow \textit{SubRec}; \\
\quad \textit{connected}' : \textit{Subs} \rightarrowtail\!\!\!\!\nrightarrow \textit{Subs} \bullet \\
\quad \textit{Free}' = \textit{Subs} \\
\quad \textit{Unavailable}' = \{\} \\
\quad \textit{call}' = \{\} \\
\quad \langle \textit{Free}', \textit{Unavailable}', \\
\qquad (\mathrm{dom}\ \textit{call}') \cup (\mathrm{ran}\ \textit{connected}')\rangle\ \textit{partition}\ \textit{Subs} \\
\quad \textit{Callers}' = \mathrm{dom}\ ((\textit{call}' \mathbin{;} \textit{st}) \rhd \textit{Connected}) \\
\quad \textit{connected}' = \textit{Callers}' \lhd (\textit{call}' \mathbin{;} \textit{num}) \\
\hline
\end{array}
$$

We can simplify this by eliminating the quantified variables, after making the necessary substitutions

$$
\begin{array}{|l}
\hline
\textit{InitExist} \\
Subs \in \mathbb{P}\, Subs \\
\{\} \in \mathbb{P}\, Subs \\
(dom\ ((\{\} \mathbin{;} st) \rhd Connected)) \in Subs \\
\{\} \in Subs \pfun SubRec \\
((dom\ ((\{\} \mathbin{;} st) \rhd Connected)) \lhd (\{\} \mathbin{;} num)) \in \\
\qquad Subs \pinj Subs \\
\langle Subs, \{\}, \\
(dom\ \{\}) \cup \\
\qquad (ran\ ((dom\ ((\{\} \mathbin{;} st) \rhd Connected)) \lhd (\{\} \mathbin{;} num)))\rangle \\
\qquad\qquad partition\ Subs \\
\hline
\end{array}
$$

Notice that

$$
\begin{aligned}
dom\ & ((\{\} \mathbin{;} st) \rhd Connected) \\
& = dom\ (\{\} \rhd Connected) \\
& = dom\ \{\} \\
& = \{\}
\end{aligned}
$$

Making this substitution in InitExist gives

$$
\begin{array}{|l}
\hline
\textit{InitExist} \\
Subs \in \mathbb{P}\, Subs \\
\{\} \in \mathbb{P}\, Subs \\
\{\} \in \mathbb{P}\, Subs \\
\{\} \in Subs \pfun SubRec \\
(\{\} \lhd \{\}) \in Subs \pinj Subs \\
\langle Subs, \{\}, (dom\ \{\}) \cup (ran\ (\{\} \lhd \{\}))\rangle\ partition\ Subs \\
\hline
\end{array}
$$

Simple properties of sets and power sets reduce this to

$$
\begin{array}{|l}
\hline
\textit{InitExist} \\
(\{\} \lhd \{\}) \in Subs \pinj Subs \\
\langle Subs, \{\}, (dom\ \{\}) \cup (ran\ (\{\} \lhd \{\}))\rangle\ partition\ Subs \\
\hline
\end{array}
$$

Domain restriction is strict with respect to the empty set

InitExist

$\{\} \in Subs \rightarrowtail\!\!\!\!\rightarrow Subs$
$\langle Subs, \{\}, (dom\ \{\}) \cup (ran\ \{\})\rangle\ partition\ Subs$

which is further simplified to

InitExist

$\langle Subs, \{\}, (dom\ \{\}) \cup (ran\ \{\})\rangle\ partition\ Subs$

The domain and range of the empty mapping are both themselves empty

InitExist

$\langle Subs, \{\}, \{\} \cup \{\}\rangle\ partition\ Subs$

and set union is idempotent

InitExist

$\langle Subs, \{\}, \{\}\rangle\ partition\ Subs$

So all that we are left with is a simple theorem about set partitions; therefore there is an initial state.

InitExist

true

QED

9.4 The Operations

Having defined the state, we should now describe the operations which are permitted upon it.

Lifting the Handset

Consider first of all what happens when a subscriber lifts the handset of a telephone which is not ringing. We shall call this operation on the exchange state *Lift*. From the subscriber's point of view, there are only two situations in which *Lift* is applicable: when the subscriber is free; and when the subscriber is party to a *suspended* telephone call. The latter requires some explanation. If s_1 calls s_2 and establishes a successful connection—we shall find out what that means in a moment—then it is s_1 and *only* s_1 who can cancel the connection. If s_2 hangs up, then the call is *suspended*; it may be re-established by s_2 lifting the handset again. Many telephone systems have this design feature—often unnoticed by the average subscriber! So specifying our operation falls into two parts.

Case 1: When the Subscriber is Free

First, if the subscriber was free before the operation then the subscriber succeeds in seizing a line

LiftFree

$$\Delta Exchange$$
$$s? : Subs$$

$$s? \in Free$$
$$Free' = Free \setminus \{s?\}$$
$$Unavailable' = Unavailable$$
$$call' = call \cup \{s? \mapsto (seize, \langle\rangle)\}$$

LiftFree is free from certain disagreeable side-effects, as is shown by Lemmas 9.1, 9.2, and 9.3.

Lemma 9.1 *The operation LiftFree doesn't alter the set of Callers*

$$LiftFree \vdash Callers' = Callers$$

Proof

$$\begin{aligned}
&Callers' \\
&\quad = \operatorname{dom}((call' \mathbin{;} st) \rhd Connected) \\
&\quad = \operatorname{dom}(((call \cup \{s? \mapsto (seize, \langle\rangle)\}) \mathbin{;} st) \rhd Connected) \\
&\quad = \operatorname{dom}(((call \mathbin{;} st) \cup \\
&\qquad\quad (\{s? \mapsto (seize, \langle\rangle)\} \mathbin{;} st)) \rhd Connected)
\end{aligned}$$

$$
\begin{aligned}
&= \mathit{dom}\ (((\mathit{call} \mathbin{;} \mathit{st}) \rhd \mathit{Connected}) \cup \\
&\qquad ((\{s? \mapsto (\mathit{seize}, \langle\rangle)\} \mathbin{;} \mathit{st}) \rhd \mathit{Connected})) \\
&= \mathit{dom}\ ((\mathit{call} \mathbin{;} \mathit{st}) \rhd \mathit{Connected}) \cup \\
&\quad \mathit{dom}\ ((\{s? \mapsto (\mathit{seize}, \langle\rangle)\} \mathbin{;} \mathit{st}) \rhd \mathit{Connected}) \\
&= \mathit{Callers} \cup \mathit{dom}\ ((\{s? \mapsto (\mathit{seize}, \langle\rangle)\} \mathbin{;} \mathit{st}) \rhd \mathit{Connected}) \\
&= \mathit{Callers} \cup \mathit{dom}\ (\{s? \mapsto \mathit{seize}\} \rhd \mathit{Connected}) \\
&= \mathit{Callers} \cup \mathit{dom}\ \{\} \\
&= \mathit{Callers} \cup \{\} \\
&= \mathit{Callers}
\end{aligned}
$$

QED

Lemma 9.2 *The operation LiftFree doesn't affect the connected subscribers*

$$\mathit{LiftFree} \vdash \mathit{connected}' = \mathit{connected}$$

Proof

$$
\begin{aligned}
&\mathit{connected}' \\
&\quad = \mathit{Callers}' \lhd \mathit{call}' \\
&\quad = \mathit{Callers} \lhd \mathit{call}' \\
&\quad = \mathit{Callers} \lhd (\mathit{call} \cup \{s? \mapsto (\mathit{seize}, \langle\rangle)\}) \\
&\quad = (\mathit{Callers} \lhd \mathit{call}) \cup (\mathit{Callers} \lhd \{s? \mapsto (\mathit{seize}, \langle\rangle)\}) \\
&\quad = \mathit{connected} \cup (\mathit{Callers} \lhd \{s? \mapsto (\mathit{seize}, \langle\rangle)\})
\end{aligned}
$$

But, from theorem 9.6

$$s? \in \mathit{Free} \Leftrightarrow s? \notin \mathit{Callers}$$

$$\mathit{connected}' = \mathit{connected} \cup \{\} = \mathit{connected}$$

QED

Lemma 9.3 *LiftFree doesn't confuse calls, so that it leaves the relation between subscribers and calls in progress (call) functional*

$$\mathit{LiftFree} \vdash \mathit{call}' \in \mathit{Subs} \pfun \mathit{SubRec}$$

Proof

First, note that by hypothesis

$$call \in Subs \rightarrowtail\!\!\!\!\!\!\rightarrow SubRec$$

and by construction

$$\{s? \mapsto (seize, \langle\rangle)\} \in Subs \rightarrowtail\!\!\!\!\!\!\rightarrow SubRec$$

since

$$s? \in Subs \land (seize, \langle\rangle) \in SubRec$$

From theorem 9.5 and by hypothesis we have

$$\begin{aligned}
& Free \cap (dom\ call) = \{\} \land s? \in Free \\
& \quad \Leftrightarrow Free \cap (dom\ call) = \{\} \land \{s?\} \subseteq Free \\
& \quad \Rightarrow \{s?\} \cap (dom\ call) = \{\} \\
& \quad \Leftrightarrow (dom\ call) \cap \{s?\} = \{\} \\
& \quad \Leftrightarrow (dom\ call) \cap (dom\ \{s? \mapsto (seize, \langle\rangle)\}) = \{\} \\
& \quad \Leftrightarrow (call \cup \{s? \mapsto (seize, \langle\rangle)\}) \in Subs \rightarrowtail\!\!\!\!\!\!\rightarrow SubRec
\end{aligned}$$

QED

Now we can prove the theorem that really interests us.

Theorem 9.8 *The precondition for LiftFree is simply that the subscriber is free. Define PreLiftFree to be the precondition for LiftFree*

$$\begin{array}{|l}
\hline
\textit{PreLiftFree} \\
\hline
\textit{Exchange} \\
s? : Subs \\
\hline
\exists\ Free', Unavailable', Callers' : \mathbb{P}\ Subs;\ call' : Subs \rightarrowtail\!\!\!\!\!\!\rightarrow SubRec; \\
\quad connected' : Subs \rightarrowtail\!\!\!\!\!\!\rightarrowtail Subs \bullet \\
\quad \langle Free', Unavailable', \\
\quad (dom\ call') \cup (ran\ connected')\rangle\ partition\ Subs \\
\quad Callers' = dom\ ((call'\ \text{;}\ st) \rhd Connected) \\
\quad connected' = Callers' \lhd (call'\ \text{;}\ num) \\
\quad s? \in Free \land Free' = Free \setminus \{s?\} \\
\quad Unavailable' = Unavailable \\
\quad call' = call \cup \{s? \mapsto (seize, \langle\rangle)\} \\
\hline
\end{array}$$

Since Free is a set of subscribers, s? is a subscriber, and set difference is total, we have

$$Free \setminus \{s?\} \in Subs$$

Thus we can eliminate Free′ from the quantified predicate in PreLiftFree

$$
\begin{array}{|l}
\hline
\textit{PreLiftFree} \\
Exchange \\
s? : Subs \\
\hline
\exists\, Unavailable', Callers' : \mathbb{P}\, Subs;\ call' : Subs \pfun SubRec; \\
\quad connected' : Subs \pinj Subs \bullet \\
\quad \langle Free \setminus \{s?\}, Unavailable', \\
\quad\ (\mathrm{dom}\ call') \cup (\mathrm{ran} connected')\rangle \text{ partition } Subs \\
\quad Callers' = \mathrm{dom}\,((call' \comp st) \rres Connected) \\
\quad connected' = Callers' \dres (call' \comp num) \\
\quad s? \in Free \land Unavailable' = Unavailable \\
\quad call' = call \cup \{s? \mapsto (seize, \langle\rangle)\} \\
\hline
\end{array}
$$

Since by definition and by Lemmas 9.1, and 9.2 we have

$$
\begin{array}{l}
Unavailable' = Unavailable \\
Callers' = Callers \\
connected' = connected
\end{array}
$$

we can eliminate Unavailable′, Callers′, and connected′

$$
\begin{array}{|l}
\hline
\textit{PreLiftFree} \\
Exchange \\
s? : Subs \\
\hline
\exists\, call' : Subs \pfun SubRec \bullet \\
\quad \langle Free \setminus \{s?\}, Unavailable, \\
\quad\ (\mathrm{dom}\ call') \cup (\mathrm{ran}\ connected)\rangle \\
\qquad \text{partition } Subs \\
\quad s? \in Free \\
\quad call' = call \cup \{s? \mapsto (seize, \langle\rangle)\} \\
\hline
\end{array}
$$

Eliminating call′ is also easy, since we have proved by Lemma 9.3 that adding our new maplet to call results in a function

$$\begin{array}{|l}
\text{__ } PreLiftFree \text{ ____________________} \\
Exchange \\
s? : Subs \\
\hline
\langle Free \setminus \{s?\}, Unavailable, \\
\quad (\mathrm{dom}\ (call \cup \{s? \mapsto (seize, \langle\rangle)\})) \cup (\mathrm{ran}\ connected)\rangle \\
\qquad \mathrm{partition}\ Subs \\
s? \in Free \\
\hline
\end{array}$$

But we have

$$\langle Free, Unavailable, (\mathrm{dom}\ call) \cup (\mathrm{ran}\ connected)\rangle\ \mathrm{partition}\ Subs$$

so by a law about partitions

$$\begin{array}{|l}
\text{__ } PreLiftFree \text{ ____________________} \\
Exchange \\
s? : Subs \\
\hline
s? \in Free \\
\hline
\end{array}$$

QED

Case 2: When the Call is Suspended

Next, if the subscriber $s?$ is party to a suspended call then someone is connected to $s?$, and that someone has status *suspended*. The subscriber that $s?$ is connected to is

$$connected^{-1}\ s?$$

The status of this caller is

$$(connected^{-1} \mathbin{;} call \mathbin{;} st)\ s?$$

Now our subscriber $s?$ is party to a suspended call if

$$(s? \mapsto suspended) \in (connected^{-1} \mathbin{;} call \mathbin{;} st)$$

In this case we simply replace the status *suspended* by *speech*

$$
\begin{array}{|l}
\hline
\textit{LiftSuspend} \\
\Delta \textit{Exchange} \\
s? : \textit{Subs} \\
\hline
(s? \mapsto \textit{suspended}) \in (\textit{connected}^{-1} \circ \textit{call} \circ \textit{st}) \\
\textit{Free}' = \textit{Free} \\
\textit{Unavailable}' = \textit{Unavailable} \\
\textit{call}' = \textit{call} \oplus \{(\textit{connected}^{-1}\ s?) \mapsto (\textit{speech}, s?)\} \\
\hline
\end{array}
$$

Exercise 9.1 *Prove the following theorem*

Theorem 9.9 *The precondition for LiftSuspend is that the subscriber is suspended.*

Our complete definition of *Lift* puts these two alternatives together

$$\textit{Lift} \ \widehat{=}\ \textit{LiftFree} \vee \textit{LiftSuspend}$$

Dialling a Digit

Now consider what happens when a subscriber dials a digit. This operation is only applicable if the subscriber has currently seized a line, or if the subscriber is already in the process of dialling. First we shall define a function which tells us what the next state is, depending on the digits dialled so far

$$
\begin{array}{|l}
\textit{nextstate} : \textit{seq}[\textit{Digit}] \rightarrow \textit{Status} \\
\hline
\forall n : \textit{seq}[\textit{Digit}] \bullet \\
\quad\quad n \in \textit{Subs} \Rightarrow \textit{nextstate}\ n = \textit{connecting} \\
\quad\quad n \in \textit{Valid} \setminus \textit{Subs} \Rightarrow \textit{nextstate}\ n = \textit{dialling} \\
\quad\quad n \notin \textit{Valid} \Rightarrow \textit{nextstate}\ n = \textit{unobtainable}
\end{array}
$$

Exercise 9.2 *Prove the following theorem*

Theorem 9.10 *The function nextstate is total. It is sufficient to prove that one of the alternatives in the definition must hold, whatever the value of n. That is,*

$$\forall n : \textit{seq}[\textit{Digit}] n \in \textit{Subs} \vee n \in \textit{Valid} \setminus \textit{Subs} \vee n \notin \textit{Valid}$$

It is easy to describe the effect of *Dial* with the benefit of *nextstate*

$$
\begin{array}{|l}
\textit{Dial} \\
\hline
\Delta \textit{Exchange} \\
s? : \textit{Subs};\ d? : \textit{Digit} \\
\hline
((s? \mapsto \textit{seize}) \in (\textit{call} \mathbin{;} \textit{st}) \lor \\
\quad (s? \mapsto \textit{dialling}) \in (\textit{call} \mathbin{;} \textit{st})) \\
\textit{Free}' = \textit{Free} \\
\textit{Unavailable}' = \textit{Unavailable} \\
\textit{call}' = \textit{call} \oplus \{s? \mapsto (\textit{nextstate}\ \textit{newnum}, \textit{newnum})\} \\
\textbf{where} \\
\quad \textit{newnum} = ((\textit{call} \mathbin{;} \textit{num})\ s?) \frown \langle d? \rangle \\
\hline
\end{array}
$$

Exercise 9.3 *Prove the following theorem*

Theorem 9.11 *The precondition for Dial is that either the subscriber has seized a line, or the subscriber is in the middle of dialling.*

Answering a Call

Lift is the operation corresponding to the initiation of a telephone call by a free user. We provide the complementary operation: *Answer*. This operation is only applicable when the subscriber in question is being called by another and the call is in the state *ringing*

$$(s? \mapsto \textit{ringing}) \in (\textit{connected}^{-1} \mathbin{;} \textit{call} \mathbin{;} \textit{st})$$

Exercise 9.4 *Prove the following theorem*

Theorem 9.12 *If a called subscriber s is being called by another and the call is ringing, then s is free.*

The only change that is made is to update the status to that of *speech*

$$\textit{call}' = \textit{call} \oplus \{(\textit{connected}^{-1}\ s?) \mapsto (\textit{speech}, s?)\}$$

The entire operation is defined as

Answer

$$\begin{array}{l}
\Delta Exchange \\
s? : Subs \\
\hline
(s? \mapsto ringing) \in (connected^{-1} \mathbin{;} call \mathbin{;} st) \\
Free' = Free \setminus \{s?\} \\
Unavailable' = Unavailable \\
call' = call \oplus \{(connected^{-1}\ s?) \mapsto (speech, s?)\}
\end{array}$$

Exercise 9.5 *Prove the following theorem*

Theorem 9.13 *The precondition for Answer is that the subscriber is party to a call which is ringing.*

Hanging up the Telephone

Finally, we must describe what happens when a subscriber hangs up the handset—*clears down* in telephony jargon. There are three cases where clearing down is applicable: if the subscriber is the initiator of a telephone call; if the subscriber is the recipient of a telephone call; and if the subscriber is unavailable. We shall consider each case separately.

Case 1: When the Subscriber is the Initiator of a Call

Suppose that $s?$ is the initiator of a telephone call; that is

$$s? \in dom\ call$$

Here there are also three cases: when $s?$ is attempting to connect (*seize*, *dialling*, *connecting*, *engaged*, or *unobtainable*); when a connection has been established, and the called party is on-hook (*ringing* or *suspended*); and when a connection has been established and the called party is off-hook (*speech*).

If $s?$ is attempting to connect, then $s?$ becomes free and the call information is discarded

$$
\begin{array}{|l}
\hline
\textit{ClearAttempt} \\
\Delta \textit{Exchange} \\
s? : \textit{Subs} \\
\hline
s? \in \textit{dom call} \\
(\textit{call} \fatsemi \textit{st})\ s? \in \\
\quad \{\textit{seize}, \textit{dialling}, \textit{connecting}, \textit{engaged}, \textit{unobtainable}\} \\
\textit{Free}' = \textit{Free} \cup \{s?\} \\
\textit{Unavailable}' = \textit{Unavailable} \\
\textit{call}' = \{s?\} \lhd \textit{call} \\
\hline
\end{array}
$$

Exercise 9.6 *Prove the following theorem*

Theorem 9.14 *The precondition for ClearAttempt is that the subscriber is making a call which has not yet been connected.*

If $s?$ is connected to another subscriber who is on-hook, then *both* subscribers are freed and the call information discarded

$$
\begin{array}{|l}
\hline
\textit{ClearLine} \\
\Delta \textit{Exchange} \\
s? : \textit{Subs} \\
\hline
s? \in \textit{dom call} \\
(\textit{call} \fatsemi \textit{st})\ s? \in \{\textit{ringing}, \textit{suspended}\} \\
\textit{Free}' = \textit{Free} \cup \{s?, \textit{connected}\ s?\} \\
\textit{Unavailable}' = \textit{Unavailable} \\
\textit{call}' = \{s?\} \lhd \textit{call} \\
\hline
\end{array}
$$

Exercise 9.7 *Prove the following theorem*

Theorem 9.15 *The precondition for ClearLine is that the subscriber is the initiator of a call which is connected, but the recipient is on-hook.*

Finally, if $s?$ is connected to another subscriber who is off-hook, then $s?$ is freed, and the other subscriber seizes a line

ClearConnect

$$\Delta Exchange$$
$$s? : Subs$$

$$s? \in \mathrm{dom}\ call$$
$$(call \mathbin{;} st)\ s? = speech$$
$$Free' = Free \cup \{s?\}$$
$$Unavailable' = Unavailable$$
$$call' = (\{s?\} \lhd call) \cup \{connected\ s? \mapsto (seize, \langle\rangle)\}$$

This is another design decision in our system.

Exercise 9.8 *Prove the following theorem*

Theorem 9.16 *The precondition for ClearConnect is that the subscriber is making a call which is connected, and the other subscriber is off-hook.*

Case 2: When the Subscriber is the Recipient of a Call

If $s?$ is the recipient of a telephone call, rather than the initiator

$$s? \in \mathrm{ran}\ connected$$

then attempting to clear down merely suspends the call

ClearSuspend

$$\Delta Exchange$$
$$s? : Subs$$

$$s? \in \mathrm{ran}\ connected$$
$$Free' = Free$$
$$Unavailable' = Unavailable$$
$$call' = call \oplus \{connected^{-1}\ s? \mapsto (suspended, s?)\}$$

Exercise 9.9 *Prove the following theorem*

Theorem 9.17 *The precondition for ClearSuspend is that the subscriber is the recipient of a connected call.*

Case 3: When the Subscriber is Unavailable

We have at last reached the significance of the set *Unavailable*: if a handset has been off-hook for too long—whatever that is—then the exchange assumes that the subscriber has forgotten to hang up and consequently considers that the subscriber is unavailable. When this happens in reality, the exchange usually sends an unpleasant tone down the line as its way of asking: "Is anybody there?". If $s?$ is unavailable

$$s? \in \mathit{Unavailable}$$

then clearing the line makes $s?$ free again

$$\begin{array}{|l}
\textit{ClearUnavail} \\
\hline
\Delta \mathit{Exchange} \\
s? : \mathit{Subs} \\
\hline
s? \in \mathit{Unavailable} \\
\mathit{Free}' = \mathit{Free} \cup \{s?\} \\
\mathit{Unavailable}' = \mathit{Unavailable} \setminus \{s?\} \\
\mathit{call}' = \mathit{call} \\
\hline
\end{array}$$

Exercise 9.10 *Prove the following theorem*

Theorem 9.18 *The precondition for the ClearUnavail operation requires that the subscriber is unavailable.*

Putting all these together gives a completed definition of *Clear*

$$\begin{aligned}
&\mathit{Clear} \mathrel{\widehat{=}} \\
&\quad \mathit{ClearAttempt} \lor \mathit{ClearLine} \lor \mathit{ClearConnect} \lor \\
&\quad \mathit{ClearSuspend} \lor \mathit{ClearUnavail}
\end{aligned}$$

Connecting Calls

This completes the description of the operations that users may initiate. The alert reader will be wondering how subscribers ever get connected together, since none of the operations described actually establishes a connection. Thinking about it, it is clear that there is another operation that is *internal* to the telephone exchange. It is not initiated by any subscriber, but rather it happens whenever the exchange feels like it, and can find a subscriber who is connecting. The effect of the operation depends on whether the called party is free or not. If it is, then it is removed from the free set and a record made that it is ringing. Otherwise it must have not been free—it was *engaged*; so this is the new status, the free set remaining unchanged.

$$
\begin{array}{|l}
\textit{Connect} \\
\hline
\Delta \textit{Exchange} \\
s? : \textit{Subs} \\
\hline
(\textit{call} \mathbin{;} \textit{st})\ s? = \textit{connecting} \\
\textit{Unavailable}' = \textit{Unavailable} \\
((s? \notin \operatorname{dom} \textit{connected} \\
\ \ (\textit{call} \mathbin{;} \textit{num})\ s? \in \textit{Free} \\
\ \ \textit{Free}' = \textit{Free} \setminus \{s?\} \\
\ \ \textit{call}' = \textit{call} \oplus \{s? \mapsto (\textit{ringing}, (\textit{call} \mathbin{;} \textit{num})\ s?)\}) \\
\lor \\
\ (s? \notin \operatorname{dom} \textit{connected} \\
\ \ (\textit{call} \mathbin{;} \textit{num})\ s? \notin \textit{Free} \\
\ \ \textit{Free}' = \textit{Free} \\
\ \ \textit{call}' = \textit{call} \oplus \{s? \mapsto (\textit{engaged}, (\textit{call} \mathbin{;} \textit{num})\ s?)\})) \\
\hline
\end{array}
$$

Exercise 9.11 *Prove the following theorem*

Theorem 9.19 *The precondition for Connect is that the subscriber has initiated a call that is connecting.*

Exercise 9.12 *There is just one more operation that should be described. What is it?*

Consequences of the Operations

A good habit to get into when writing specifications is to investigate properties of our system. Examples of important properties that our telephone system had better possess are given in the following theorems.

Exercise 9.13 *Prove the following theorems*

Theorem 9.20 *A subscriber who succeeds in connecting to his or her own number will discover that it is engaged.*

Theorem 9.21 *A subscriber who succeeds in connecting to his or her own number is able to clear the call.*

Theorem 9.22 *The system must not* deadlock*: it must not reach a configuration of the state in which no further operation can take place.*

Summary of Operations

A summary of the telephone exchange state is given in Figure 9.1, and of the operations on the state in Figure 9.2

$$
\begin{array}{|l}
\textit{Exchange} \\
\hline
\mathit{Free}, \mathit{Unavailable}, \mathit{Callers} : \mathbb{P}\, \mathit{Subs} \\
\mathit{call} : \mathit{Subs} \nrightarrow \mathit{SubRec} \\
\mathit{connected} : \mathit{Subs} \rightarrowtail\!\!\!\!\nrightarrow \mathit{Subs} \\
\hline
\langle \mathit{Free}, \mathit{Unavailable}, (\mathrm{dom}\ \mathit{call}) \cup (\mathrm{ran}\ \mathit{connected}) \rangle \\
\quad \mathrm{partition}\ \mathit{Subs} \\
\mathit{Callers} = \mathrm{dom}\ ((\mathit{call} \mathbin{;} \mathit{st}) \rhd \mathit{Connected}) \\
\mathit{connected} = \mathit{Callers} \lhd (\mathit{call} \mathbin{;} \mathit{num}) \\
\hline
\end{array}
$$

Figure 9.1: The Telephone Exchange State

Operation	Input	Precondition
LiftFree	$s? : Subs$	$s? \in Free$
LiftSuspend	$s? : Subs$	$(s? \mapsto suspended) \in (connected^{-1} \fatsemi call \fatsemi st)$
Dial	$s? : Subs$ $d? : Digit$	$(s? \mapsto seize) \in (call \fatsemi st) \lor$ $(s? \mapsto dialling) \in (call \fatsemi st)$
Answer	$s? : Subs$	$(s? \mapsto ringing) \in (connected^{-1} \fatsemi call \fatsemi st)$
ClearAttempt	$s? : Subs$	$s? \in \mathrm{dom}\ call$ $(call \fatsemi st)\ s? \in \{seize, dialling, engaged, connecting, unobtainable\}$
ClearLine	$s? : Subs$	$s? \in \mathrm{dom}\ call$ $(call \fatsemi st)\ s? \in \{ringing, suspended\}$
ClearConnect	$s? : Subs$	$s? \in \mathrm{dom}\ call$ $(call \fatsemi st)\ s? = speech$
ClearSuspend	$s? : Subs$	$s? \in \mathrm{ran}\ connected$
ClearUnavail	$s? : Subs$	$s? \in Unavailable$
Connect		$(call \fatsemi st)\ s? = connecting$

Figure 9.2: Preconditions for the Telephone Exchange Operations

Chapter 10

Case Study: A Theory of Clocks

An important advantage of using mathematics is that we can extend our notation whenever we wish, so that a particular specification becomes easier to write and thereafter to read. This chapter is devoted to the development of a small piece of theory which is then exploited in a formal specification. The mathematics is by no means deep, but it does give interesting insights into the problem in hand. The message is this: if we develop specific mathematical theories for the systems that we wish to specify, we can gain a much deeper understanding of those systems, as well as better descriptions.

10.1 A Revolutionary Theory

In this section we shall develop a modest theory of clock revolutions[1]. First of all, we discuss the positions on a clock's face, and draw some consequences of the definitions that we propose; next, we describe some operators for manipulating clock positions.

[1]This theory owes much to the comments of Steve Powell, a colleague at IBM UK Laboratories, Hursley Park.

Positions on a Clock Face

A clock is an instrument for measuring time. Clocks come in many different guises, but the kind that we have in mind consists of a number of *positions* arranged in a circle, with one or more *hands* pointing to them. The hands can be distinguished, and, for the sake of this disposition, each hand has one end fixed at the centre of the clock. The positions sometimes carry markings, such as spots, lines, or numerals, to help the accuracy of readings—but only *sometimes*. Fashion and æsthetics both have a part to play in the design of timepieces.

Since the positions on a clock are arranged in a circle, it is convenient to choose one position as being distinguished. Each position may be referred to by the angle at the centre of the clock through which a hand must travel to get from the distinguished point to the chosen position. Let the number of divisions on our clock be denoted by the symbol $\odot$. We shall not define precisely what this is, but we feel that there should be at least one division

$$\begin{array}{|l}\odot : \mathbb{N} \\ \hline \odot \neq 0\end{array}$$

On a clock we consider the angle subtended by a hand from the distinguished position $\odot$—the angle from 12 o'clock—to come from the set

$$Angle \mathrel{\widehat{=}} 0 \mathinner{..} (\odot - 1)$$

The fact that the hand may have gone through several complete revolutions from its starting point before reaching its current position is regarded as unimportant. Thus we consider certain positions for a hand on the clock to be *congruent* to each other: a position a and a position b are congruent if their difference is an exact number of whole revolutions. In practical terms, if we see a hand on a conventional clock pointing to the number "3", we can easily see that the angle subtended from 12 o'clock is a right angle. It is irrelevant whether it has travelled through a hundred or a thousand and one revolutions, and then one right angle to get to its present position. The important fact is that it is one right angle from 12 o'clock. All "the same positions on the face" form a congruence class, and we can define these congruence classes as follows

$$
\begin{array}{|l}
[\_] : \mathbb{Z} \to \mathbb{P}\,\mathbb{Z} \\
\hline
\forall k : \mathbb{Z} \bullet \\
\qquad [k] = \{i : \mathbb{Z} \bullet k + i \times \odot\}
\end{array}
$$

where $\mathbb{Z}$ is the set of all integers. So

$$[k]$$

denotes the class of all integers congruent to k *modulo* $\odot$. For example, for an angle a

$$[a] = \{\ldots, a - 2 \times \odot, a - \odot, a, a + \odot, a + 2 \times \odot, \ldots\}$$

Consequences of the Definition

We can think of several consequences of the definitions in the last section. Every position on the clock face is uniquely determined by an angle; in other words each congruence class contains precisely one angle (Theorem 10.1). To prove this, we first show that each congruence class is nonempty (Lemma 10.1), and then show that each congruence class contains at most one angle (Lemma 10.2).

Lemma 10.1 *Every congruence class is nonempty, in particular*

$$\forall k : \mathbb{Z} \bullet k \in [k]$$

Proof

1	$0 \times \odot = 0$	property of $\times$
2	$k = k + 0$	property of $+$
3	$k = k + 0 \times \odot$	2, 1 subst
4	$\exists i : \mathbb{Z} \bullet k = k + i \times \odot$	3 $\exists$-introduction
5	$k \in \{i : \mathbb{Z} \bullet k + i \times \odot\}$	4 comprehension
6	$k \in [k]$	by definition

QED

Lemma 10.2 *Each congruence class contains at most one angle*

$$\forall k : \mathbb{Z} \bullet \forall a, b : Angle \cap [k] \bullet a = b$$

Proof

Suppose that there were more than one angle in a congruence class $[k]$*; call them* a *and* b*, where*

$$a \neq b$$

From this supposition we would derive a contradiction.

$$\begin{aligned}
& a \in Angle \cap [k] \\
& \quad \Leftrightarrow a \in Angle \land a \in [k] \\
& \quad \Leftrightarrow a \in Angle \land \exists i : \mathbb{Z} \bullet a = k + i \times \odot \\
& \quad \Leftrightarrow \exists i : \mathbb{Z} \bullet a \in Angle \land a = k + i \times \odot \\
& \quad \Leftrightarrow \exists i : \mathbb{Z} \bullet (k + i \times \odot) \in Angle \\
& \quad \Leftrightarrow \exists i : \mathbb{Z} \bullet 0 \leq (k + i \times \odot) \leq \odot - 1
\end{aligned}$$

Similarly

$$\begin{aligned}
& b \in Angle \cap [k] \\
& \quad \Leftrightarrow \exists j : \mathbb{Z} \bullet 0 \leq (k + j \times \odot) \leq \odot - 1
\end{aligned}$$

In order for a *and* b *to be distinct, the values of* i *and* j *in the two quantified formulæ must also be distinct. Without any loss of generality, we can choose an* i *and a* j *such that*

$$i > j$$

The i *and* j *chosen must satisfy the inequalities*

$$\begin{aligned}
& 0 \leq (k + i \times \odot) \leq \odot - 1 \\
& 0 \leq (k + j \times \odot) \leq \odot - 1
\end{aligned}$$

Now, if x *and* y *are two positive numbers each less than or equal to* z*, then a simple property of numbers says that the difference* $x - y$ *is* *also less than z. Thus we can deduce that*

$$
\begin{aligned}
&(0 \leq (k + i \times \odot) \leq \odot - 1) \wedge (0 \leq (k + j \times \odot) \leq \odot - 1) \\
&\quad \Rightarrow (k + i \times \odot) - (k + j \times \odot) \leq \odot - 1 \\
&\quad \Leftrightarrow k + i \times \odot - k - (j \times \odot) \leq \odot - 1 \\
&\quad \Leftrightarrow i \times \odot - (j \times \odot) \leq \odot - 1 \\
&\quad \Leftrightarrow (i - j) \times \odot \leq \odot - 1
\end{aligned}
$$

Now, consider i and j once more

$$i - j > 0 \Leftrightarrow i - j \geq 1$$

Since $\odot$ is strictly positive, we can multiply both sides of this last inequality by $\odot$, giving

$$(i - j) \times \odot \geq \odot$$

But this is a contradiction, since we have just proved that

$$(i - j) \times \odot \leq \odot - 1$$

Therefore our assumption that a and b were distinct was wrong, so

$$a = b$$

QED

Theorem 10.1 *Each class contains precisely one angle*

$$\vdash \ \forall k : \mathbb{Z} \bullet \#(Angle \cap [k]) = 1$$

Proof: From Lemmas 10.1 and 10.2.

The last theorem showed that we can measure every position on a clock's face by an angle. Later we shall see that the hands on a clock arrive at their positions by travelling perhaps through many revolutions. Our next theorem will show that a hand can only be in one position on a clock's face at a time.

Theorem 10.2

$$[x] = [k] \Leftrightarrow x \in [k]$$

Proof

Suppose that $[x] = [k]$

$$x \in [x] \Leftrightarrow x \in [k]$$

Therefore

$$[x] = [k] \Rightarrow x \in [k]$$

Suppose that $x \in [k]$

$$x \in [k] \Leftrightarrow \exists i : \mathbb{Z} \bullet x = k + i \times \odot$$

For any $t : \mathbb{Z}$,

$$\begin{aligned} & t \in [x] \\ & \quad \Leftrightarrow \exists j : \mathbb{Z} \bullet t = x + j \times \odot \\ & \quad \Leftrightarrow \exists i : \mathbb{Z} \bullet x = k + i \times \odot \wedge \exists j : \mathbb{Z} \bullet t = x + j \times \odot \\ & \quad \Leftrightarrow \exists i, j : \mathbb{Z} \bullet t = (k + i \times \odot) + j \times \odot \\ & \quad \Leftrightarrow \exists i, j : \mathbb{Z} \bullet t = k + (i + j) \times \odot \\ & \quad \Leftrightarrow \exists l : \mathbb{Z} \bullet t = k + l \times \odot \\ & \quad \Leftrightarrow t \in [k] \end{aligned}$$

Hence membership of $[x]$ *and membership of* $[k]$ *is the same thing*

$$\forall t : \mathbb{Z} \bullet t \in [x] \Leftrightarrow t \in [k]$$

Therefore

$$[x] = [k]$$

and so

$$x \in [k] \Rightarrow [x] = [k]$$

Now we have proved our implications both ways, so

$$[x] = [k] \Leftrightarrow x \in [k]$$

QED

Two Operators

There are two natural operations that yield angles, because any operation that produces an integer now determines a unique angle. Our two operations correspond to addition and multiplication

$$
\begin{array}{|l}
\_\oplus\_ : \mathbb{Z} \times \mathbb{Z} \rightarrow Angle \\
\_\otimes\_ : \mathbb{Z} \times \mathbb{Z} \rightarrow Angle \\
\hline
\forall m, n : \mathbb{Z} \bullet \\
\qquad m \oplus n \in [m + n] \\
\qquad m \otimes n \in [m \times n]
\end{array}
$$

The fact that these are uniquely defined follows from Theorem 10.1.

Exercise 10.1 *Prove the following laws, some of which we have given names so that we can refer to them later*

$$
\begin{array}{ll}
x \oplus y \in Angle & \\
(x \oplus y) \oplus z = x \oplus (y \oplus z) & \\
x \oplus y = y \oplus x & \\
x \oplus 0 = x & \\
x \oplus (-x) = 0 & \\
x \otimes y \in Angle & \\
(x \otimes y) \otimes z = x \otimes (y \otimes z) & \otimes \text{ associative} \\
x \otimes y = y \otimes x & \\
0 \otimes x = 0 & \text{zero for } \otimes \\
x \otimes 1 = x & \\
x \otimes (y \oplus z) = (x \otimes y) \oplus (x \otimes z) & \\
(x \oplus y) \otimes z = (x \otimes z) \oplus (y \otimes z) & \otimes \text{ back dist over } \oplus
\end{array}
$$

Finally, a law which is like a kind of cancellation law

$$a, b, c : Angle \vdash a \oplus b = a \oplus c \Leftrightarrow b = c$$

The fact that both b and c are angles allows us to equate them without worrying about discounting revolutions. A half rotation is defined as

$$
\begin{array}{|l}
\delta : Angle \\
\hline
\delta \neq 0 \\
\delta \otimes 2 = 0
\end{array}
$$

10.2 A Clock

We can describe an analogue clock as a pair of angles measured from 12 o'clock. One angle describes the position of the hour hand ($h : Angle$) and the other, the minute hand ($m : Angle$). But a clock is more than just these two hands: they *must* be in a particular relationship to one another. In fact, if we were to hide away the minute hand, we could tell the time from the hour hand alone. It should be easy to see what this relationship is: for every division on the clock that the hour hand advances, the minute hand moves twelve times as far

$$m = h \otimes 12$$

where $\otimes$ was defined in the last section. Thus, our notion of a clock consists of a declaration and a predicate, which together form a schema which we shall call *Clock*

```
┌─ Clock ─────────────────────────
│ h, m : Angle
├─────────────
│ m = h ⊗ 12
└─────────────────────────────────
```

10.3 Manipulating Clocks

The following operation inverts a clock, that is, it rotates the clock through half a revolution

```
┌─ Invert ────────────────────────
│ Δ Clock
├─────────────
│ h' = h ⊕ δ
│ m' = m ⊕ δ
└─────────────────────────────────
```

The reflection of a clock through its vertical axis is defined by the following operation

```
┌─ ReflectV ──────────────────────────────
│ ΔClock
├──────────
│ h' ⊕ h = 0
│ m' ⊕ m = 0
└─────────────────────────────────────────
```

10.4 A Proof Obligation

One theorem which we must prove about an operation that we define on a system state, is that it is *applicable*. In order that we really can invert a clock, we must prove that

$$Clock \vdash \exists\, Clock' \bullet Invert$$

In fact this is not a theorem; to prove this we derive a contradiction

1	$\exists\, Clock' \bullet Invert$	by hypothesis
2	$\exists\, Clock' \bullet h' = h \oplus \delta \wedge m' = m \oplus \delta$	by definition
3	$\exists\, h', m' : Angle \bullet$ $m' = h' \otimes 12 \wedge h' = h \oplus \delta \wedge$ $m' = m \oplus \delta$	by definition
4	$\exists\, h', m' : Angle \bullet m \oplus \delta = (h \oplus \delta) \otimes 12$	by substitution
5	$m \oplus \delta = (h \oplus \delta) \otimes 12$	by elimination
6	$m = h \otimes 12 \wedge m \oplus \delta = (h \oplus \delta) \otimes 12$	by hypothesis
7	$(h \otimes 12) \oplus \delta = (h \oplus \delta) \otimes 12$	by substitution
8	$(h \otimes 12) \oplus \delta = (h \otimes 12) \oplus (\delta \otimes 12)$	$\otimes$ back dist over $\oplus$
9	$\delta = \delta \otimes 12$	cancellation law
10	$\delta = (\delta \otimes 2) \otimes 6$	$\otimes$ associativity
11	$\delta = 0 \otimes 6$	definition of δ
12	$\delta = 0$	zero for $\otimes$

which contradicts the definition of δ. Thus there are no states satisfying the precondition for *Invert*. This means that there are no circumstances in which we can invert a clock: it never makes sense to do so!

Surprisingly, perhaps, we have failed to prove the applicability of the clock inversion operation. This demonstrates the usefulness of a simple proof: we now know that we can never invert a clock and end up with a feasible result. If our system development relied on us providing an implementation of *Invert*, then we have just prevented further wasted effort.

10.5 More Clocks

Consider the following object

```
┌─ SmallClock ──────────────────────
│ hand : Angle
└───────────────────────────────────
```

with the operation

```
┌─ Tick ────────────────────────────
│ ΔSmallClock
├───────────
│ hand' = hand ⊕ 1
└───────────────────────────────────
```

Tick is of course total.

Clocks actually come with more than one hand; you might expect to see this clock around the house

```
┌─ HouseClock ──────────────────────
│ SmallClock[min/hand]
│ SmallClock[hr/hand]
├───────────
│ min = hr ⊗ 12
└───────────────────────────────────
```

where $SmallClock[min/hand]$ is just the same as *SmallClock*, but with *min* systematically substituted for *hand*. We would expect a house clock to tick at least once a minute

```
┌─ MinTick ─────────────────────────
│ ΔHouseClock
│ Tick[min/hand]
└───────────────────────────────────
```

Before we proceed, we should check the precondition of this operation, so that our clock may run.

Exercise 10.2 *What is the precondition for MinTick? Under what circumstances is this possible?*

A happier clock is defined as

$$\begin{array}{|l} \hline \textit{HourTick} \\ \Delta \textit{Clock} \\ \textit{Tick}[\textit{hr}/\textit{hand}] \\ \hline \end{array}$$

Exercise 10.3 *What is the precondition for HourTick?*

We can define a general rotation of the clock as

$$\begin{array}{|l} \hline \textit{Rotate} \\ \Delta \textit{HouseClock} \\ a? : \textit{Angle} \\ \hline hr' = hr \oplus a? \\ min? = min \oplus a? \\ \hline \end{array}$$

representing a rotation through a given angle $a?$.

Exercise 10.4 *Investigate the precondition of Rotate. Are any rotations possible? If not, why not? If there are, how many?*

We next consider the most general reflection. This is not determined by the *Angle* of the line about which we reflect, for that would exclude perfectly good reflections that don't correspond to *Angles*. Rather, the appropriate generalisation is to determine the reflection by providing the *Angle to which a hand at zero would be reflected.*

The general reflection of a clock is given by

$$\begin{array}{|l} \hline \textit{HandReflect} \\ \Delta \textit{SmallClock} \\ a? : \textit{Angle} \\ \hline hand' \oplus hand = a? \\ \hline \end{array}$$

where $a?$ is the angle of the line which would be the reflection of a hand at angle zero. The angle that the original hand makes with the zero position is the same as the angle that the new hand makes with the $a?$ position, but in the opposite sense—draw a diagram! This leads to

$$hand = a? \oplus (-hand')$$

which can be rewritten as above. Let

$$\begin{array}{|l}
\textit{Reflect} \\
\hline
\textit{HandReflect}[min / hand, min' / hand'] \\
\textit{HandReflect}[hr / hand, hr' / hand'] \\
\hline
\end{array}$$

Exercise 10.5 *Investigate the precondition of Reflect.*

10.6 Particular Clocks

Exercise 10.6 *Clocks are particularly interesting things, as the following few exercises attempt to show.*

1. *If* $\odot = 11$, *how does the HouseClock behave?*
2. *If* $\odot = 13$, *how does the HouseClock behave?*
3. *If* $\odot = 60$, *how does the HouseClock behave?*
4. *What is a* reasonable *value for* $\odot$*?*

Chapter 11

Algebras

In the previous chapters of this book we have built a number of theories, some describing general mathematical objects and some acting as specifications of simple systems. The strategy we have adopted for theory building has been to enrich existing bodies of theory, thus sequences, for example, were modelled in terms of functions (and hence in terms of relations and sets). The specifications of systems that were developed modelled aspects of the system, such as its state and operations, by appropriate configurations of sets, relations, functions and sequences, consequently this approach to specification is sometimes referred to as *model based.* In this chapter we will introduce an alternative style for presenting theories which uses only the underlying logical system, and does not explicitly introduce general mathematical objects for the purposes of modelling. The underlying logical system we will use is a restricted form of predicate logic with equality.

We will start by introducing a theory of natural numbers, which up to now we have taken for granted. This will be followed by a theory of sequences, which describes precisely the same mathematical objects as were discussed in Chapter 8, but without modelling them as special types of functions. Adopting this approach we can clearly see how the *theory of sequences*, for example, can be distinguished from particular *representations of sequences.* We will also develop a small example of a system specification, to show how this style of theory presentation, sometimes called *algebraic specification*, can be applied to software engineering.

Mathematical models, or representations, which satisfy a theory are called *algebras* of the theory. Mathematicians have studied algebras as objects in their own right for many years, and we will conclude the chapter by introducing some of the ideas and terminology necessary to make accessible the literature related to this study.

11.1 Equational Theory Presentations

The underlying logical system we are going to use for the theory presentations in this chapter makes use of universally quantified equations. There are no predicates other than equality, and the semantics we will give to equations is slightly different from that introduced earlier in the book. The substitution rule will be interpreted in contexts outside of propositions, allowing us to discuss *values* of objects rather than just the truth of statements. We will say that the substitution of equal terms into an expression allows the expression to be *rewritten* to an equivalent expression. Consequently the axioms of such a system are sometimes referred to as *rewrite rules.* If we wish to state that an expression, $\mathcal{A}$, can be rewritten to another expression, $\mathcal{B}$, by the application of one or more of the rewrite rules given as axioms of the system, we will write

$$\mathcal{A} \leadsto \mathcal{B}$$

We will now build a theory of the natural numbers. Just as building the theory of sets involved describing the rules governing the operations we can perform on objects which are instances of sets, so building our theory of natural numbers involves describing not the symbols $\{1, 2, 3, \ldots\}$, but the *collection of operations* which can be carried out on the objects which the symbols denote. In fact, because we are not using a model based approach, we do not need to consider how to represent the natural numbers in any way other than in terms of the operations upon them. We will refer to our theory of natural numbers as a specification of the *data type*[1] *natural.*

[1] You should be aware that the terms *type, data type* and *abstract data type* are used in a variety of different ways in software engineering, frequently without explanation.

Some of our theories will need to refer to values of different *sorts*. The expression $x > y$, for example, might form part of our theory of the data type *natural*, but it will have a truth value and not a numerical one. It is important, therefore, that we state what sorts of values our variables denote in quantified expressions. You must remember to distinguish between the *data type natural*, the *set* $\mathbb{N}$, which is the collection of all possible values of a particular sort in the data type, and "4" which is a possible representation of one value of the sort.

We will consider a very simple version of the data type *natural*, one in which the numbers can only be added together. For this theory we need only one sort, which we will call NAT. To build up the theory presentation, we must choose the terms in our formal system so that we expose some of the structure implicit in the ways values of the data type *natural* can be constructed. Observe that every number, except 0, can be thought of as the one after some previous number, so it can be named the "successor of x" where x is the name of the previous number; we will abbreviate this to $s(x)$. It should be obvious that the intended interpretation of this expression is the function

$$\lambda x : \mathbb{N} \bullet x + 1$$

but because we are working before any interpretation we will refer to it as a *unary operation* which has as its *signature*[2]

$$s : NAT \to NAT$$

We also need a binary operation to represent addition, which we will present here in prefix form $add(x, y)$ although it usually interprets to a function in infix form such as $x + y$. It has as its signature

$$add : NAT\ NAT \to NAT$$

The special natural 0 we will represent by *zero* in the theory. This can be viewed as a *constant* or *nullary operation*, having no arguments,

[2]In the algebraic approach, the distinction between total and partial functions is not made in the signature, consequently the notation $f : A \to B$ should be read as indicating that a mathematical interpretation of f is a total or partial function from A to B.

whose signature is

$$zero : \to NAT$$

All we have done so far is to present the signature of the theory, by giving the signatures of the operations. When applied to valid arguments, all of these operations return values of the sort NAT, and so they provide different ways of naming values of the sort. Next we describe the properties of these operations, by using the predicate for equality to say when different terms are describing the same value. We will present these as equalities between terms, where any variable symbols are assumed to be universally quantified over all values of the appropriate sort.

First we will observe that adding *zero* to any number leaves its value unchanged

$$add(zero, x) = x$$

Next we will say that adding $s(x)$ to y is the same as adding x to $s(y)$. This is just a statement of the fact that $(x + 1) + y = x + (y + 1)$.

$$add(s(x), y) = add(x, s(y))$$

This is all we need to say in the theory presentation as all the other properties we want can be derived as theorems. It is useful to present all aspects of the theory in a common framework, so Figure 11.1 shows it in its entirety, expressed in a suitable metalanguage.

We can now make use of our axioms as a set of rewrite rules. To perform a rewrite correctly we have only to identify an appropriate rule to apply. For example, we can see that

$$add(s(s(s(zero))), s(s(zero))) \rightsquigarrow add(s(s(zero)), s(s(s(zero))))$$

by observing that it is of the form $add(s(x), y)$.

Exercise 11.1 *Show that*

$$add(s(s(zero)), s(zero)) \rightsquigarrow s(s(s(zero)))$$

Data Type: *natural*
Sorts: NAT
Operations:
$zero : \to NAT$
$s : NAT \to NAT$
$add : NAT\ NAT \to NAT$
Variables:
$x : NAT$
Equations:
$add(zero, x) = x$
$add(s(x), y) = add(x, s(y))$
End Data Type.

Figure 11.1: The Data Type *natural*.

Turning to our intended model for the theory, we must assign meanings to every expression that can be constructed in the theory, in such a way that our axioms interpret to true equalities. First we will make the global observation that our sort NAT will correspond to the set $\mathbb{N}$. This means that every term of the data type which is of this sort (in this case, every term), must be given as its interpretation a member of the set {0,1,2,...}. It should not come as any surprise that we will use the following mapping

$$\{zero \mapsto 0,\ s(zero) \mapsto 1,\ s(s(zero)) \mapsto 2,\ \cdots\}$$

All terms of the form $add(x, y)$ can be rewritten to precisely one which is of the form $s(s(\cdots(zero)\cdots))$. The interpretation we will give to $add(x, y)$ is the meaning of the term it rewrites to. We can now prove simple arithmetic results, such as $3 + 2 = 5$.

Exercise 11.2 *Show that*

$$add(s(s(s(zero))), s(s(zero))) \leadsto s(s(s(s(s(zero)))))$$

Remember that the interpretation we have just given for this theory is only one possible model; there are many others, some of which may involve choosing different symbols to represent the values of the sort NAT, such as

$$\{0,1,10,11,100,101,\ldots\} \qquad \text{or} \qquad \{I, II, III, IV, V, VI,\ldots\}$$

Others may not be so obvious. It is the ability to find a variety of models for such theories that forms the basis of computational systems. Specification can be thought of as producing a theory which describes the artifact to be built, and design can be thought of as finding the most appropriate model of the theory.

Exercise 11.3 *Show that the set* $\{x : \mathbb{N} \bullet 2x\}$*, together with the operation of addition, is a model for the theory of naturals.*

One model which is so obvious that it is often overlooked is called the *term algebra*. This consists of the strings of symbols that we are using to denote terms in the theory, so that, for example, the term *zero* has as its interpretation the string "zero". The rewrite rules are then interpreted as simple string substitution rules. This means we can manipulate the textual representation of the theory as a model in its own right.

Unfortunately there are usually a number of models for a theory which we do not really want. Consider using the set $\{v\}$, containing just one element, to model the sort NAT, so that every term in the theory must map onto this one value.

$$\{zero \mapsto v,\ s(zero) \mapsto v,\ s(s(zero)) \mapsto v,\ \cdots\}$$

If we look at the equations for the theory we will find that they all hold true, and this is indeed a model for the theory, although not a particularly useful one. The problem is that our theory presentation only tells us when things are equal, and not when they are different. There is nothing to stop us accepting models where $s(x) = zero$, even though we have not stated that this should be the case. This property of an interpretation is referred to as *confusion*.

If we use the set $\mathbb{N} \cup \{v\}$ as our interpretation of the sort and use the mapping

$$\{zero \mapsto 0, s(zero) \mapsto 1, s(s(zero)) \mapsto 2, \cdots\}$$

as before then another problem arises. We have one value, v, which does not correspond to a term in the theory. This will not violate any of the equations in the theory since nothing can be said for this rogue value to violate! Values such as these are sometimes referred to as *junk*.

There are many advantages to restricting our attention to models which have *no junk and no confusion* – such models are called *initial*. It must be stressed that this initiality is not a property of the theory, but of the interpretation that it is being given; it is being given an *initial semantics*. In what follows we will only consider initial models of theories.

It is very important to be able to re-use simple theory presentations, as we do not want to keep re-specifying wheels! We avoid having to repeat the presentation of a theory by providing, in the metalanguage, the power to enrich existing theories of data types to make new ones. We can create a theory of modulo-3 arithmetic, for example, by noting that it shares many of the properties of our ordinary arithmetic for naturals. Our new theory needs just one additional rule:

$$s(s(s(zero))) = zero$$

We will write this enrichment as

Data Type: *natural modulo* 3 enrichment of *natural*
Equations:
$$s(s(s(zero))) = zero$$
End Data Type.

This means that the theory for naturals is to be carried forward, with just one new equation added, to form a theory for a *new data type*[3]. We will consider enrichment as a purely textural device, replicating the sorts, operations and equations of one theory in another.

[3]Care needs to be taken to ensure that the resulting theory is consistent: it is very easy to add new rules which contradict those of the theory being enriched. For some purposes in artificial intelligence, where learning systems are being discussed, formal systems which capture the addition of contradictory axioms are used. These are known as *non-monotonic logics*.

The exact semantics of enrichment, however, are far more complex than this. You should be aware, for example, that enrichment could cause multiple instances of one theory to be included in another, and that we need to consider whether these instances must all be given the same interpretation or not. We have managed to avoid this problem by using enrichment only in very simple ways, and we will continue to do so.

Exercise 11.4 *Use this enriched theory to show that* $2 + 2 = 1$.

Exercise 11.5 *Enrich the theory of natural numbers to build a theory of natural numbers which includes the notion of multiplication.*

Exercise 11.6 *Write a theory of integers that will admit the notions of addition, subtraction and unary minus. Enrich this to include multiplication.*

11.2 Heterogeneous Theories

Let us consider another theory, that describing sequences of items. If SEQ is a sort in our theory which will interpret to sequences of items, then we have a number of operations such as

$$\begin{array}{l} reverse : SEQ \rightarrow SEQ \\ join : SEQ\ SEQ \rightarrow SEQ \\ tail : SEQ \rightarrow SEQ \end{array}$$

which are called *closed* operations since they take arguments of one sort and will return values of the same sort. We also need other operations, however, which are not closed since they involve more than one sort of value, such as

$$\begin{array}{l} addi : ITEM\ SEQ \rightarrow SEQ \\ length : SEQ \rightarrow NAT \\ head : SEQ \rightarrow ITEM \end{array}$$

Theories of this kind are called *heterogeneous* or *many-sorted.*

Here is part of a theory presentation which has *as one of its models* sequences as introduced in Chapter 8. We will assume that the sequences are sequences of values of some previously defined type, called *item*. We will also make use of the type *natural* as already defined, but we will use an infix form of the add operator, +, and the symbol 1 to denote $s(zero)$ as this makes the presentation easier to read.

Data Type: *sequence* enrichment of *natural* and *item*

Sorts: *SEQ*

Operations:

$emptys : \to SEQ$

$length : SEQ \to NAT$

$addi : ITEM\ SEQ \to SEQ$

$join : SEQ\ SEQ \to SEQ$

$tail : SEQ \to SEQ$

$head : SEQ \to ITEM$

Variables:

$x : NAT$

$s : SEQ$

Equations:

$length(emptys) = zero$

$length(addi(i, s)) = 1 + length(s)$

$head(addi(i, s)) = i$

$tail(addi(i, s)) = s$

$join(emptys, t) = t$

$join(addi(i, s), t) = addi(i, join(s, t))$

End Data Type.

You will notice here that we have to pay close attention to the patterns we match in the rules. The operations length and join are rewritten in different ways for the cases

$$s = emptys \qquad \text{and} \qquad s \neq emptys$$

the second case being expressed as the fact that s is formed by adding at least one element to a sequence, since all non-empty sequences can be rewritten to this form.

Example 11.1 *Here we will show that*

$$join(addi(x, addi(y, emptys)), addi(z, emptys))$$
$$\rightsquigarrow addi(x, addi(y, addi(z, emptys)))$$

Rewrites:

$$join(addi(x, addi(y, emptys)), addi(z, emptys))$$
$$\rightsquigarrow addi(x, join(addi(y, emptys), addi(z, emptys)))$$
$$\rightsquigarrow addi(x, addi(y, join(emptys, addi(z, emptys))))$$
$$\rightsquigarrow addi(x, addi(y, addi(z, emptys)))$$

QED

Exercise 11.7 *Enrich the theory of sequences given above to include all the operations introduced in Chapter 8, then use your rewrite rules to show the following:*

1. $\#(\langle s, t\rangle \frown \langle u, v, w\rangle) = \#\langle s, t\rangle + \#\langle u, v, w\rangle$
2. $\langle p, q, r, s, t\rangle\ 3 = r$
3. $\#rev\ \langle s, t, u, v\rangle = \#\langle s, t, u, v\rangle$
4. $rev(\langle a, b, c\rangle \frown \langle d, e\rangle) = (rev\langle d, e\rangle) \frown (rev\langle a, b, c\rangle)$

One worry you might have at this point is that we have not said how to rewrite expressions arising from taking the head or the tail of the empty sequence. The reason for this is that you are not allowed to do it, so we do not feel obliged to tell you what will happen if you do! This is akin to a physicist's refusal to say what happens in the theory of relativity if you exceed the speed of light. Unfortunately some programmers do not feel obliged to check that they are working within the theory of the types they are using, and some specifiers of types feel obliged to cater for this strange behaviour by introducing error types to handle this abuse – we prefer to encourage users of our theories to work within them.

Some operations will return a value which we want to interpret to "true" or "false" in a model. These operations correspond to the predicates that we are not allowed to use explicitly in the theory presentation, so what we have done is to convert them to functions which return values of some data type which will be interpreted as true or false. It is no different to any other type we might want to use, and so we need to give the corresponding theory for it. If we never want to do anything with values of this type but write and read them then all we need is

Data Type: *truth values*

Sorts: *BOOL*

Operations:

$T : \to BOOL$

$F : \to BOOL$

End Data Type.

Note that our stated intention to use only initial models is important here, as otherwise we might encounter perverse individuals who allow both T and F to denote the same truth value, and this would certainly lead to confusion.

If we want to manipulate these values with operations corresponding to the connectives than we must enrich this theory to include the required operators; this data type is usually called *boolean* and is shown in Figure 11.2.

Exercise 11.8 *Show that*

$$and(not(T), or(not(F), not(F))) \leadsto F$$

Exercise 11.9 *Enrich this theory to include operations corresponding to $\Rightarrow$ and $\Leftrightarrow$, then explore some of the derivations of Chapter 2 using your rewrite rules.*

We can now add statements to our theories which correspond to predicates other than equality, by introducing new boolean valued operations. Here is a data type which is an enrichment of *NAT* and *BOOL* which adds the operation $even(x)$. This will evaluate to T if x is even, and F if x is odd, and hence captures the same notion as that of a predicate.

Data Type: *boolean* enrichment of *truth values*

Operations:

$not : BOOL \to BOOL$

$and : BOOL\ BOOL \to BOOL$

$or : BOOL\ BOOL \to BOOL$

Variables:

$b : BOOL$

Equations:

$not(T) = F$

$not(F) = T$

$and(F, b) = F$

$and(T, b) = b$

$or(T, b) = T$

$or(F, b) = b$

End Data Type.

Figure 11.2: The Data Type *boolean*

Data Type: *natural even* enrichment of *natural* and *boolean*

Operations:

$even : NAT \to BOOL$

Variables:

$x : NAT$

Equations:

$even(zero) = T$

$even(s(zero)) = F$

$even(s(s(x))) = even(x)$

End Data Type.

Exercise 11.10 *Enrich the theory of sequences to include an operation which determines if its argument is the empty sequence.*

Exercise 11.11 *Enrich the theory of naturals to include an operation for telling whether one natural is greater than or equal to another.*

11.3 The Theory of a High–Low Store

In this section we will demonstrate the use of these techniques by specifying a small system. The system is a component of a real-time control system. It has the task of recording readings from a number of identical, named, sensors. It must respond to a number of requests for information, providing the following

- the maximum reading for a named device.
- the minimum reading for a named device.
- whether or not any device has recorded the greatest reading a sensor can register. This is deemed an alarm value.
- whether a given device has any readings recorded.
- the maximum and minimum readings recorded by each device.

We need a number of sorts of values associated with this task. First we need a sort for the store itself, which we will call HLS. Second we need sorts for the names and readings of the sensors: we will call these SN and SR respectively. Finally we need BOOL to represent the answers to the questions which are truth valued.

With the benefit of this notation we can now write down the signatures of the operations corresponding to the functions of the store.

First we observe that we need an empty store from which all others can be formed.

$$newhls : \rightarrow HLS$$

Next we need an operation to add readings to the store, or, more precisely, to create a new store out of a sensor name, a sensor reading and an old store.

$$insertsr : SN\ SR\ HLS \rightarrow HLS$$

We also need the operations which will return the maximum and minimum readings recorded for a named device.

$$maxreading : SN\ HLS \rightarrow SR$$
$$minreading : SN\ HLS \rightarrow SR$$

The operations to return the absolute maximum and minimum of all the sensor readings recorded do not need a sensor name.

$$absmaxreading : HLS \rightarrow SR$$
$$absminreading : HLS \rightarrow SR$$

The operation to determine if the alarm should sound, because a reading has been recorded of the greatest that a sensor can register, and the operation to see if a particular sensor has recorded any readings, are both truth valued.

$$alarm : HLS \rightarrow BOOL$$
$$readings : SN\ HLS \rightarrow BOOL$$

We have already given the theory presentation for a data type *boolean* which is sufficient for our purposes. The data type in which the sensor names are defined is trivial, since all we need is a collection of names: there are no operations that we need to perform with these names. In fact, we could leave the specification *loose* by omitting the specification of this data type, or *parameterised* by expressing it in such a way that any suitable data type can be used to provide the theory of names. To enable rewrites to be shown in a definite way, however, we have included a specification of a data type *sensor name* in Figure 11.3 which captures a world in which there are just three sensors.

Similar choices are open to us concerning the data type *sensor reading*, except that we need access to the operations in this data type that enable us to compare values of the sort *SR*, so that we can establish minimum and maximum values. We will develop such a theory.

Each sensor provides a reading on a discrete scale with seven possible readings. We do not need to worry about how these values

Data Type: *Sensor Name*

Sorts: *SN*

Operations:

$a : \rightarrow SN$

$b : \rightarrow SN$

$c : \rightarrow SN$

End Data Type.

Figure 11.3: The Data Type *sensor names*

are represented, of course, or whether they are evenly spaced, as our sole concerns are to be able to order them and to detect any alarm readings. There are two special sensor readings[4].

$smallestsr : \rightarrow SR$
$largestsr : \rightarrow SR$

All other sensor readings can be related to the smallest.

$nextsr : SR \rightarrow SR$

Finally we must be able to compare two sensor readings to find out which is the smaller and which is the larger.

$smallersr : SR\ SR \rightarrow SR$
$largersr : SR\ SR \rightarrow SR$

The equations governing these operations are recorded in a complete data type specification, in Figure 11.4 on page 258. You should note the way in which we have ensured that no values greater than the greatest value can be recorded.

We can now return to our task of writing the equations for our data type ***high low store***. We will use variables $d, d1, d2$ to denote

[4]Our decision to include *largestsr* enables us to develop a data type for our high low store in which the store does not need to know how many different values the sensor is capable of registering.

sensor device names, r to denote a reading, and s to denote a store. The equations for discovering whether a particular sensor has recorded a reading are easy to produce. If the store is empty then clearly no readings have been recorded for any sensor.

$$readings(d, newhls) = F$$

If the store under consideration was constructed by adding a sensor reading for the named device then clearly there is a reading

$$readings(d, insertsr(d, r, s)) = T$$

but if the reading was added from another sensor then we must investigate further.

$$readings(d1, insertsr(d2, r, s)) = readings(d1, s)$$

The operation to establish whether any device has recorded the largest sensor reading possible is defined as follows.

$$alarm(newhls) = F$$
$$alarm(insertsr(d, largestsr, s)) = T$$
$$alarm(insertsr(d, r, s)) = alarm(s) \ IF \ r \neq largestsr$$

We have guarded the third equation, by adding a conditional construct, to ensure that it never matches the same case as the second equation.

The equations for *absmaxreading* need to make use of the operation *largersr* to compare sensor readings. We need to decide whether asking for the absolute maximum reading in an empty store should be a defined operation. We will decide that it should not be defined[5]. For a store having only one value inserted clearly this value is the maximum in the store.

$$absmaxreading(insertsr(d, r, newhls)) = r$$

[5]This is only reasonable if we have provided a mechanism by which a system using this data type can establish if any readings are present. As we have provided the *readings* operation, we feel justified in requiring any users to establish if there are any readings before trying to request the maximum.

For a store with more than one value, we must compare the last one inserted with the absolute maximum of all the rest.

$$absmaxreading(insertsr(d, r, s)) = largersr(r, absmaxreading(s))$$

Specifying the operation to retrieve the largest sensor reading stored by a particular sensor is facilitated by introducing an auxiliary operation. This is purely a device for easing the task of presenting the theory, and it must be made clear to anyone reading the specification that there is no requirement for this operation to be implemented. Certainly this operation should not be made available to users of the data type. The operation we require will take a store and a sensor name, and return a value which is a store restricted to contain only the entries for the named sensor.

$$restrict : \ SN \ \ HLS \rightarrow HLS$$

The following equations will serve to define this operation.

$$restrict(d, newhls) = newhls$$
$$restrict(d1, insertsr(d1, r, s)) = insertsr(d1, r, restrict(d1, s))$$
$$restrict(d1, insertsr(d2, r, s)) = restrict(d1, s)$$

Now *maxreading* is easy to specify.

$$maxreading(d, s) = absmaxreading(restrict(d, s))$$

Note that the onus is still on the user to check that there is at least one reading, this time for the named sensor, or the result will be undefined.

Unfortunately we have no corresponding operation *absminreading* to facilitate the specification of *minreading*, so we will introduce one as an auxiliary equation. The full specification of the data type *high low store* is presented in Figure 11.5 on page 259.

Exercise 11.12 *Rewrite the following expressions to their simplest forms.*

1. *readings*$(a,$ *insertsr*$(a,$ *nextsr*(*nextsr*(*smallestsr*)), *newhls*))
2. *absmaxreading*(*insertsr*$(a,$ *nextsr*(*nextsr*(*smallestsr*)), *newhls*))
3. *minreading*$(a,$ *insertsr*$(a,$ *nextsr*(*nextsr*(*smallestsr*)), *newhls*))
4. *alarm*(*insertsr*$(a,$ *largestsr*, *newhls*))
5. *maxreading*$(a,$ *insertsr*$(a,$ *nextsr*(*nextsr*(*smallestsr*)), *insertsr*$(a,$ *nextsr*(*smallestsr*), *newhls*)))
6. *maxreading*$(a,$ *insertsr*$(a,$ *nextsr*(*nextsr*(*smallestsr*)), *newhls*))

Exercise 11.13 *Develop algebraic specifications for the storage management systems and the configuration management system specified earlier in the book.*

11.4 Algebras

In this section we are going to introduce some of the mathematics which is used to study models of equational theories. We will restrict our attention to single-sorted theories, since that is where most of the mathematical ideas originated. A mathematical model of an equational theory has two parts

- a set, $\mathcal{A}$, of symbols denoting values of the type. This set is referred to as the *carrier set* of the algebra.

- a set, Ω, of closed operations on the carrier set, possibly including nullary, constant, operations.

The carrier set and the operations are collectively referred to as an *algebra* or *algebraic system*. This is shown by pairing together the carrier and the set of operations in square brackets, thus $[\mathcal{A}, \Omega]$. Where no confusion can arise, we will drop the set brackets around the operations. The algebra which models our natural numbers with addition and multiplication is

$$[\mathbb{N}, \{0, +, \times\}] \qquad \text{or} \qquad [\mathbb{N}, 0, +, \times]$$

There are a number of levels at which algebras can be treated, ranging from a study of the whole concept of an algebra to studies of particular sets and operations. We are going to look at classes of simple algebras which satisfy certain axioms in their theories, and also some concrete examples of algebras of these classes.

We will begin by looking at algebras which have only one binary in-fix operation, $\star$ say. Such algebras are called *groupoids* and must satisfy the following theory.

Data Type: *groupoid*
Sorts: *VALUE*
Operations:
$\_\star\_ : VALUE\ VALUE \rightarrow VALUE$
End Data Type.

Models of this theory form a very general class of algebras since the only restriction is that there must be just one closed binary operation, and no other properties are required.

Exercise 11.14 *Which of the following are not groupoids, and why?*

1. $[\mathbb{N}, \times]$
2. $[\mathbb{N}, -]$
3. $[\mathbb{P}\,\mathbb{N}, \cup]$
4. $[\mathbb{Z}, -]$
5. $[(\mathbb{P}\,\mathbb{N}) \setminus \{\}, \cap]$
6. $[F_{N \rightarrow N}, {}_{9}^{\circ}]$ *where* $F_{N \rightarrow N}$ *denotes the set of all* $\mathbb{N} \rightarrow \mathbb{N}$ *functions.*

The next class of algebras satisfies the additional restriction that the operation is *associative*. Such algebras are called *semigroups* and their theory can be presented as an enrichment of that for groupoids.

Data Type: *semigroup* enrichment of *groupoid*

Variables:

$x, y, z : VALUE$

Equations:

$x \star (y \star z) = (x \star y) \star z$

End Data Type.

Exercise 11.15 *Which of the algebras in Exercise 11.14 are not semigroups and why?*

Members of the next class of algebras have a special element, in addition to the binary operation, so we add a nullary operation and equations to define its properties. Algebras which are models of this theory are called *monoids*.

Data Type: *monoid* enrichment of *semigroup*

Operations:

$i : \rightarrow VALUE$

Variables:

$x : VALUE$

Equations:

$i \star x = x$

$x \star i = x$

End Data Type.

This special element i is referred to as an *identity* for the operation.

Example 11.2 $[\mathbb{N}, 1, \times]$ *is a monoid, since* $[\mathbb{N}, \times]$ *is a semigroup and* 1 *is an identity for multiplication.*

Exercise 11.16 *Find the operations marked "?" which would make the following monoids.*

1. $[\mathbb{N}, +, ?]$
2. $[\mathbb{Z}, -, ?]$
3. $[\mathbb{P}\,\mathbb{N}, ?, \{\}]$

4. $[\mathsf{P}\,\mathsf{N}, \cap, ?]$

5. $[F_{N\to N}, \circ_9, ?]$ *where* $F_{N\to N}$ *denotes the set of all* $\mathsf{N} \to \mathsf{N}$ *functions.*

If we had only the first or second equation in the above enrichment then we would have a *left identity* or *right identity* respectively, but then the resulting algebras would not be monoids, of course.

We can show an interesting property of monoids, namely that if we enrich the theory to admit another identity constant, then we can prove as a theorem that the two identities are equal.

Theorem 11.1 *The following enrichment of the data type monoid has as a theorem* $\vdash i = j$.

Data Type: *trial* enrichment of *monoid*

Operations:

$j : \to VALUE$

Equations:

$j \star x = x$

$x \star j = x$

End Data Type.

Proof

Since the equations must hold for all values of the appropriate sort, including the particular ones i *and* j, *we can write down,*

$i \star j = i$ *and* $i \star j = j$

but this gives us, substituting for $i \star j$,

$i = j$

QED

This is an important result, as it lies at the heart of solving equations such as

$7 + x = 7$ and $3y = 3$

where we derive the *unique* answers of $x = 0$ and $y = 1$ respectively. We could not guarantee this uniqueness if $[\mathbb{N}, +, 0]$ and $[\mathbb{N}, *, 1]$ were not monoids.

The next enrichment we are going to make to our theory gives rise to the class of algebras called *groups*. It requires the algebra to have a unary operation which satisfies the following

> Data Type: *group* enrichment of *monoid*
> Operations:
> $\_^{-1} : VALUE \to VALUE$
> Variables:
> $x : VALUE$
> Equations:
> $x \star (x^{-1}) = i$
> $(x^{-1}) \star x = i$
> End Data Type.

This operation is referred to as an *inverse* for the operation $\star$.

Exercise 11.17 *Which of the following algebras can be groups with the selection of appropriate identities and inverses?*

1. *The set of all integers with addition.*
2. *The set of all real numbers with multiplication.*
3. *The set of all positive real numbers with multiplication.*
4. *The set of all bijective functions from A to A.*
5. *The power set of some set A with intersection.*

There is an important theorem of groups which enables the solution of equations of the form $a \star b = a \star c$.

Theorem 11.2 (Cancellation) *In a group* $[G, \star, \_^{-1}, i]$

$$\forall a, b, c : G \bullet (a \star b = a \star c) \vee (b \star a = c \star a) \Leftrightarrow b = c.$$

Exercise 11.18 *Prove the above theorem.*

We might have attempted to enrich the theory of monoids slightly differently, in order to provide different left and right inverses. For example,

Data Type: *trial* enrichment of *monoid*

Operations:

$\_^{-l} : VALUE \rightarrow VALUE$

$\_^{-r} : VALUE \rightarrow VALUE$

Variables:

$x : VALUE$

Equations:

$x \star (x^{-r}) = i$

$(x^{-l}) \star x = i$

End Data Type.

The following theorem shows that this would have been pointless.

Theorem 11.3 *The data type trial above has as a theorem*

$\vdash x^{-l} = x^{-r}$

Proof

$$\begin{aligned} x^{-l} &= x^{-l} \star i \\ &= x^{-l} \star (x \star x^{-r}) \\ &= (x^{-l} \star x) \star x^{-r} \\ &= i \star x^{-r} \\ &= x^{-r} \end{aligned}$$

QED

There is a subclass of groups where the operation is commutative, and such groups are called *abelian*.

Data Type: *abelian group* enrichment of *group*

Variables:

$x, y : VALUE$

Equations:

$x \star y = y \star x$

End Data Type.

Exercise 11.19 *Which of the groups you identified in Exercise 11.17 are abelian?*

Example 11.3 *Find the interpretation for i and $\_^{-1}$ which will make $[\{a, b, c, d\}, \star, i, \_^{-1}]$ a group, where $\star$ is defined by the following table.*

$\star$	a	b	c	d
a	a	b	c	d
b	b	c	d	a
c	c	d	a	b
d	d	a	b	c

for example, $c \star b = d$

Here we can see quite easily that i must be interpreted as a. The inverse function can be found by seeing what combinations of values give rise to the identity.

$$\_^{-1} = \{a \mapsto a, b \mapsto d, c \mapsto c, d \mapsto b\}$$

Exercise 11.20 *Find another operation, $\circ$, which is different from $\star$, and thus has a different table defining it, so that $[\{a, b, c, d\}, \circ, i, \_^{-1}]$ is a group, and find appropriate values for $\_^{-1}$ and i.*

Exercise 11.21 *If an equilateral triangle is lying on a flat surface, there are six possible transformations we can apply to it which leave it visually unchanged. These are the three rotations, through 120°, 240° and 360°, in the plane of the surface, and the three reflections about the axes of symmetry of the triangle. If we label the vertices of the triangle a, b and c, then the position of the triangle can be denoted by a triple of such values. The transformations can then be viewed as functions on the set of such triples, and sequences of transformations as compositions of such functions.*

1. *Write down the six possible functions in terms of the mappings from position to position.*

2. *Show that these functions, under the binary operation of composition, form a group, and find the inverse operation and identity.*

3. *Repeat this exercise with a square rather than a triangle.*

Now let us look at classes of algebras which have two binary operations. The first one is called a *ring* and can be built by enriching an abelian group, adding two new operations, $\circ$ and j, such that

- these operations together with the carrier set form a monoid.
- the new operation $\circ$ distributes through the operation, $\star$, of the abelian group.

Data Type: *ring* enrichment of *abelian group*

Operations:

$\circ : VALUE\ VALUE \rightarrow VALUE$

$j : \rightarrow VALUE$

Variables:

$x, y, z : VALUE$

Equations:

$x \circ (y \circ z) = (x \circ y) \circ z$

$j \circ x = x$

$x \circ j = x$

$x \circ (y \star z) = (x \circ y) \star (x \circ z)$

$(y \star z) \circ x = (y \circ x) \star (z \circ x)$

End Data Type.

Rings are of special interest because both real and integer arithmetic with operations for multiplication and addition are rings, as is arithmetic modulo-n.

Exercise 11.22 *Show that the arithmetics of integers and modulo-3 are both rings.*

Because of this close relation between ordinary arithmetic and rings, it is quite common to express the properties of rings with operator symbols that remind us of arithmetic, so that a ring $[A, +, -, 0, \cdot, 1]$ is made up from

- an abelian group $[A, +, -, 0]$
- a monoid $[A, \cdot, 1]$

- the distribution rules

$$x \cdot (y + z) = (x \cdot y) + (x \cdot z)$$

$$(x + y) \cdot z = (x \cdot z) + (y \cdot z)$$

There are a number of simple properties of arithmetic which are actually properties of all rings.

Exercise 11.23 *Prove that the following are theorems of the theory of rings.*

1. $x \cdot 0 = 0 \cdot x = 0$
2. $x \cdot (-y) = (-x) \cdot y = -x \cdot y$
3. $(-x) \cdot (-y) = x \cdot y$

There is one final classification of algebras that we will consider, which is fundamental to computer science, *boolean algebras*. We will give its full theory presentation in Figure 11.6 on page 260, to show that it is no different in nature from any of the other types of algebra we have presented.

The Boolean algebras we are most familiar with have just two elements in their carrier set.

Exercise 11.24 *Show that* $[\{T, F\}, \wedge, \vee, \neg, T, F]$ *is a Boolean algebra.*

There is, however, at least one Boolean algebra we have met in this book which admits more than two elements in the carrier set.

Exercise 11.25 *Show that* $[\mathsf{P}\, A, \cap, \cup, \{\}, A, -]$ *is a Boolean algebra, where* A *is any set and* $-$ *is the unary operation whose effect on an argument* x *is* $A \backslash x$.

11.5 Morphisms

We will conclude this chapter with a brief introduction to the "algebra of algebras". In particular, we will look at some useful relations between algebras. We will continue to restrict our attention to homogeneous algebras, and furthermore we will not consider algebras which have more than one operation of each arity. The ideas will extend to heterogeneous algebras with arbitrary numbers of operations of each arity, but the complex notation we would need to introduce to handle this more general case is not appropriate in this book. You are urged to remember, however, that these restrictions are a matter of convenience and not theoretically necessary.

The motivation behind this section is a desire to establish some idea of what it might mean for two algebras to be "equal". Clearly we will expect two algebras to be equal if we cannot tell their representations apart, that is they are identical in every respect, but are there any other forms of equivalence that exist between algebras?

We will start with a very weak form of equivalence, which can be established merely from the signatures of two algebras. We will call two homogeneous algebras *similar* if they have the same number of operations of each arity. This means, for example, that any two groups are similar, but that a group is not similar to a monoid as the latter has no unary operation. It says nothing, however, about the properties of the algebras as given in the equations: this we will do by defining what it means for two algebras to be *homomorphic*.

If two algebras $[A, \Omega]$ and $[A', \Omega']$ are similar then we can put the carrier sets and operations into some form of one-to-one correspondence. This is usually shown by the order in which the operations are written in the algebras. Assuming this has been done, then a function $\phi : A \rightarrow A'$ is a *homomorphism* from $[A, \Omega]$ to $[A', \Omega']$ if, for every $\omega \in \Omega$ and corresponding $\omega' \in \Omega'$ and every $a_i \in A$,

$$\phi(\omega(a_1, a_2, \ldots, a_n)) = \omega'(\phi(a_1), \phi(a_2), \ldots, \phi(a_n))$$

where n is established by the arity of the particular ω.

Homomorphic algebras are much more strongly related than algebras which are just similar because they are related by a relation which is *structure-preserving*. This means that the result of applying operations of Ω to values of A and then finding the images of the result under the function ϕ, is the same as the result of finding the images of the values then using the corresponding operations of Ω'. Although the function is strictly a mapping between carrier sets, it is usually referred to as a homomorphism from one algebra to another, as some knowledge of the correspondence between operations is understood.

Example 11.4 *Consider the two groupoids*

$$X = [\{T, F\}, \wedge] \qquad \textit{and} \qquad Y = [\{T, F\}, \vee]$$

These are clearly similar, and there is only one operation in each algebra so the correspondence is obvious. The function

$$\neg : \{T, F\} \rightarrowtail\!\!\twoheadrightarrow \{T, F\}$$
$$\neg = \{T \mapsto F, F \mapsto T\}$$

is a homomorphism since

$$\forall x, y : BOOL \bullet \neg(x \wedge y) = \neg x \vee \neg y$$

by DeMorgan's law.

Example 11.5 *The two algebras*

$$N = [\mathbb{N}, +, 0] \qquad \textit{and} \qquad N_m = [0 \mathinner{..} m-1, +_m, 0]$$

where $+_m$ denotes addition modulo-m, are similar. The function

$$\mathrm{mod}_m(n) : \mathbb{N} \rightarrow 0 \mathinner{..} m-1$$
$$\mathrm{mod}_m(n) = n \bmod m$$

is a homomorphism since for all $a, b \in \mathbb{N}$

$$\mathrm{mod}_m(a+b) = \mathrm{mod}_m(a) +_m \mathrm{mod}_m(b)$$

The proof of this is left as an exercise for the reader.

Example 11.6 *The two algebras*

$$R = [\mathbb{R}^+, \times] \qquad \textit{and} \qquad L = [\mathbb{R}, +]$$

are similar, and the function

$$\phi(x) : \mathbb{R}^+ \rightarrow \mathbb{R}$$
$$\phi(x) = \log(x)$$

is a homomorphism since

$$\log(x \times y) = \log(x) + \log(y)$$

This homomorphism forms the basis of the use of logarithms for allowing multiplication to be performed by looking up logarithms and adding. Both R and L are more usually thought of as groups, with the extra operations for inverses and identities. If we had presented them in this way then we would have had to show that the homomorphism had the structure preserving property for all three operations, mapping inverses to inverses and identities to identities. In fact, for groups this is simplified by the following theorem.

Theorem 11.4 *If* $[G, \star, i, \_^{-1_1}]$ *and* $[G', \star', j, \_^{-1_2}]$ *are groups and* $\phi : G \rightarrow G'$ *satisfies*

$$\forall x : G \bullet \phi(x \star y) = \phi(x) \star' \phi(y)$$

then

$$\forall x : G \bullet \phi(x^{-1_1}) = \phi(x)^{-1_2} \qquad \text{and} \qquad \phi(i) = j$$

Proof
Since i *is an identity for* $\star$

$$\phi(i) = \phi(i \star i) = \phi(i) \star' \phi(i)$$

using the given property. But $\phi(i) \in G'$ *and* j *is the identity for* $\star'$, *so*

$$j \star' \phi(i) = \phi(i)$$

and hence

$$j \star' \phi(i) = \phi(i) \star' \phi(i)$$

and so, by the Cancellation Theorem,

$$j = \phi(i)$$

and ϕ *preserves identities. We can now use this result to show that inverses are preserved too.*

$$\phi(x) \star' \phi(x^{-1_1}) = \phi(x \star x^{-1_1}) = \phi(i) = j$$

and so

$$\phi(x^{-1_1}) = \phi(x)^{-1_2}$$

QED

Thus to show that ϕ is a homomorphism between groups it is sufficient to show that the structure of the binary operation is preserved.

There are a number of special kinds of morphisms and, as is in the nature of mathematicians, they have all been given impressive sounding names. The classification is based on properties of the function ϕ. A homomorphism from $[A, \Omega]$ to $[A', \Omega']$ is

- an *isomorphism* if $\phi : A \rightarrowtail\!\!\!\!\twoheadrightarrow A'$ is bijective
- a *monomorphism* if $\phi : A \rightarrowtail A'$ is injective
- an *epimorphism* if $\phi : A \twoheadrightarrow A'$ is surjective
- an *endomorphism* if $\operatorname{ran} \phi = A$
- an *automorphism* if $\phi : A \rightarrowtail\!\!\!\!\twoheadrightarrow A$ is bijective.

Exercise 11.26 *Classify the homomorphisms given in the above examples.*

Returning to the ideas introduced earlier regarding theories and their models, we can now state an important property of initial models of a theory.

> All initial models of a theory are isomorphic to each other: in particular they are all isomorphic to the term algebra.

This is a formalisation of the fact that there are a number of ways of implementing a particular data type, all of which are equally good with respect to the theory, although they will have other properties which will differentiate them, such as performance.

11.6 Summary

In this chapter we have revisited the idea of a theory and we have shown how a theory can be given a semantics in terms of sets, functions, relations and sequences. This has been demonstrated by building theories of natural numbers and sequences. We have also explored the use of equations as rewrite rules.

We have introduced some of the terminology and ideas behind the algebraic specification of data types. Interpretations of these specifications have been discussed, and the notion of initiality has been introduced. A small case study has been developed, showing the specification of a storage system without the adoption of a model posed in terms of sets.

An introduction to the topic of algebra was presented, and classifications for homogeneous algebras were given, together with some theorems associated with common algebraic structures. The notion of morphism was discussed, and a classification of homomorphisms was given.

Data Type: *sensor reading*

Sorts: SR

Operations:

$smallestsr : \to SR$

$largestsr : \to SR$

$nextsr : SR \to SR$

$smallersr : SR\ SR \to SR$

$largersr : SR\ SR \to SR$

Variables:

$r, r1, r2 : SR$

Equations:

$$nextsr(nextsr(nextsr(nextsr(nextsr(nextsr(smallestsr))))))$$
$$= largestsr$$

$nextsr(largestsr) = largestsr$

$largersr(r, smallestsr) = r$

$largersr(smallestsr, r) = r$

$largersr(largestsr, r) = largestsr$

$largersr(r, largestsr) = largestsr$

$largersr(nextsr(r1), nextsr(r2)) = nextsr(largersr(r1, r2))$

$smallersr(r, smallestsr) = smallestsr$

$smallersr(smallestsr, r) = smallestsr$

$smallersr(nextsr(r1), nextsr(r2)) = nextsr(smallersr(r1, r2))$

End Data Type.

Figure 11.4: The Data Type *sensor reading*

Data Type: *high low store* enrichment of

sensor name and *sensor reading* and *boolean*

Sorts: SYS

Operations:

$newhls : \rightarrow SYS$

$insertsr : SN\ SR\ SYS \rightarrow SYS$

$maxreading : SN\ SYS \rightarrow SR$

$minreading : SN\ SYS \rightarrow SR$

$absmaxreading : SYS \rightarrow SR$

$alarm : SYS \rightarrow BOOL$

$readings : SN\ SYS \rightarrow BOOL$

Hidden Operations:

$restrict : SN\ SYS \rightarrow SYS$

$absminreading : SYS \rightarrow SR$

Variables:

$r, r1, r2 : SR$

$s : SYS$

$d, d1, d2 : SN$

Equations:

$readings(d, newhls) = F$

$readings(d, insertsr(d, r, s)) = T$

$readings(d1, insertsr(d2, r, s)) = readings(d1, s)$

$alarm(newhls) = F$

$alarm(insertsr(d, r, s)) = alarm(s) IF not r == largestsr$

$alarm(insertsr(d, largestsr, s)) = T$

$absmaxreading(insertsr(d, r, newhls)) = r$

$absmaxreading(insertsr(d, r, s)) = largersr(r, absmaxreading(s))$

$absminreading(insertsr(d, r, newhls)) = r$

$absminreading(insertsr(d, r, s)) = smallersr(r, absmaxreading(s))$

$restrict(d, newhls) = newhls$

$restrict(d1, insertsr(d1, r, s)) = insertsr(d1, r, restrict(d1, s))$

$restrict(d1, insertsr(d1, r, s)) = restrict(d1, s)$

$maxreading(d, s) = absmaxreading(restrict(d, s))$

$minreading(d, s) = absminreading(restrict(d, s))$

End Data Type.

Figure 11.5: The Data Type *high low store*

Data Type: *boolean algebra*

Operations:

$\star : VALUE\ VALUE \rightarrow VALUE$

$\circ : VALUE\ VALUE \rightarrow VALUE$

$i : \rightarrow VALUE$

$j : \rightarrow VALUE$

$\_^{-1} : VALUE \rightarrow VALUE$

Variables:

$x, y, z : VALUE$

Equations:

$x \star y = y \star x$

$x \circ y = y \circ x$

$x \star (y \star z) = (x \star y) \star z$

$x \circ (y \circ z) = (x \circ y) \circ z$

$i \star x = i$

$i \circ x = x$

$j \star x = x$

$j \circ x = j$

$x \star (y \circ z) = (x \star y) \circ (x \star z)$

$x \circ (y \star z) = (x \circ y) \star (x \circ z)$

$x^{-1} \star x = i$

$x^{-1} \circ x = j$

End Data Type.

Figure 11.6: The Data Type *boolean algebra*

Chapter 12

Formal Methods

In this final chapter we hope to round off the proceedings by introducing the notion of "formal methods" and showing that these are really little more than the sensible use of mathematics to assist in the process of software development. To illustrate this we will introduce some of the current generation of formal methods, and suggest the sorts of problems they might be used for.

12.1 What Are Formal Methods?

The name "formal methods" is actually a rather misleading one in some senses, since it is usually used as a classification of things which are not "methods" at all, and certainly not methods where the sequences of actions undertaken is formally defined! Rather the term has come to be used for the class of formal systems which have been designed to be useful specifically for the development of complex systems, together with the associated manuals and courses which give rise to guidelines for the formal system's use on real problems. A moment's reflection on the natures of formal systems and complex problems will show that no one formal system is ever likely to be suitable for describing and analysing *all* aspects of a complex system, so a "formal method" is not *the* method which a system designer might choose to use when developing a system, but *one of the tools* that a designer might wish to make use of during the process.

Once the rôle of a formal method as a tool has been appreciated, it is worth reminding ourselves that tools are always open to abuse. Formal methods are no exception, they can be used for the wrong tasks and in an unskilled way, and doubtlessly the users will blame the tools for the resulting shoddy workmanship.

The major rôle of this book has been to try and prepare you for the safe use of such tools, but do remember that no specific "formal method" has been explained in detail. Rather we have concentrated on the ideas that underpin the whole class of methods, such as sets, functions, and relations, and of course, on the fundamental notion of a formal system. Unfortunately, although the concepts we have introduced may be common to many methods the notation certainly is not. The tuning of the various methods towards particular tasks has lead to their designers adopting different notations. If you have understood the basic ideas, however, switching syntax from time to time should be an inconvenience rather than a major stumbling block.

12.2 Model-Oriented Specifications

One way to specify sequential systems is to use the so-called *Model-Oriented* or *State-Based* approach. The idea is to build an abstract mathematical model of the state of the system which is to be developed, together with the operations which constitute the interface. A more detailed model is then produced; this new model is nearer to the implementation and therefore less abstract. As other equally surprised authors have pointed out, this more detailed description, clouded as it is by implementation details, is called a *refinement* of the specification.

It is vitally important that the relationship between a specification and its refinement is documented. Depending on the particular method, there will be various proof obligations that must be discharged when a designer makes such a design step.

In making a refinement step, the data structures in the state may be made more concrete and the operations on the state may be decomposed into simpler operations. This step-by-step process is finished when the state and the operations are described entirely in terms of constructs available in the programming language.

The most prominent examples of the model-oriented approach are VDM[1] and Z[2]. Most of our notation is similar to that of Z. One of Z's distinctive features is the use of the structuring mechanism known as the *schema*. Although VDM[3] and Z are really very similar notations, the most striking difference is the lack of any real structuring mechanism in VDM. As a comparison between the two notations, consider the storage allocator example from Section 7.4.

First of all, we need to introduce our global constants. There are n consecutively numbered blocks

$$n : \mathbf{N}$$
$$B = \{1, \ldots, n\}$$

The notation $\{1, \ldots, n\}$ corresponds to our earlier $1..n$ notation. The *Report* type is defined as

$$Report = \{okay, fail\}$$

The two elements introduced in this manner are understood to be distinct.

The state contains a directory which tells us which blocks are owned by which users, and a set of free blocks. In VDM we would model these as a *mapping* from blocks to users and a set of blocks

$$\begin{aligned} SM :: \; & dir : \mathsf{map}\ B\ \mathsf{to}\ U \\ & free : \mathsf{set\ of}\ B \end{aligned}$$

In fact, this state is not quite the same one that we saw in Section 7.4. In VDM, there is a distinction drawn between *functions* and *maps*, which is the same as that drawn between a *partial* function and a *finite* one. Jones draws an analogy between building a map and building a table of pairs: application of a map, we are told, requires looking things up in a table rather than evaluating a defining expression.

---

[1] C.B. Jones, *Systematic Software Development Using VDM*, Prentice-Hall, 1986.

[2] I. Hayes (editor), *Specification Case Studies*, Prentice-Hall, 1987.

[3] Strictly speaking VDM is not a notation, but rather a "method" of software development: the *Vienna Development Method*. Actually, the "method" involved is usually called stepwise refinement; VDM describes the proof obligations which arise during stepwise refinement, and how they may be discharged. A real method for software development would tell its user *exactly what to do next to progress the development*. The notation used in VDM is known as META IV.

Functions are seen to be defined by a fixed rule, whereas maps are often "dynamically created". Whether or not this consideration of implementation details is a good idea, knowing what we do about functions from Chapter 7 allows us to understand the mathematics involved.

Furthermore, the power set constructor in VDM is what we have described as the *finite* power set constructor. The rationale given by Jones for finiteness in the construction is that computer stores are finite.

We recorded the invariant relationship between state items below the line in the schema describing the state. In VDM the invariant is recorded separately as a function from the state to truth values

$$\mathit{inv\text{-}SM}(\mathit{mk\text{-}SM}(\mathit{dir}, \mathit{free})) \quad \triangleq \quad \mathit{free} = B - (\mathsf{dom}\ \mathit{dir})$$

The function *mk-SM* is an injection into the type *SM*; it "tags" the pair $(\mathit{dir}, \mathit{free})$ to indicate that together they form an element of the type *SM*. In VDM the notion of types is very much more elaborate than that presented in this book. There is a hierarchy of types and their subtypes, and to prevent the problem of things belonging to more than one type, every type is supplied with an injection into the type, and the corresponding projections from the type.

The *initial* object of type *SM* is described by

$$SM_0 = \mathit{mk\text{-}SM}(\{\ \}, B)$$

We can now describe the operations on the state. *Request* is an operation which takes a user name as input and produces a block name and a report as output; it has *write access* to both of the state items. This information is recorded in the operation header

$$\begin{array}{l} \mathit{Request}\ \ (u\colon U)\ b\colon B, r\colon \mathit{Report} \\ \mathsf{ext\ wr}\ \mathit{dir}\ \ :\ \mathsf{map}\ B\ \mathsf{to}\ U \\ \quad\ \ \mathsf{wr}\ \mathit{free}\ :\ \mathsf{set\ of}\ B \end{array}$$

Each operation may have *read* access, *write* access, or *no* access to each state variable. If an operation has read access, it amounts to conjoining to the postcondition the predicate that the value of that variable is unchanged by the operation ($\mathit{cf}\ \Xi SM$).

In VDM $\overleftarrow{free}$ denotes the value of the set of free blocks before the operation *Request*, and *free* denotes its value after the operation.

In VDM, an operation is specified by providing an explicit precondition, and an explicit postcondition. The precondition is of course a predicate on the state before the operation. The postcondition[4] is a predicate on the state before and the state after the operation. Splitting the relation describing the state transition into two parts is regarded as a good discipline in VDM: it highlights the distinction between the assumption that an implementor is allowed to make, and the obligation that must be met. It also leads to some pleasing formulations of the proof obligations themselves.

It may also seem that separating the precondition from the relation has another great benefit: we need not *calculate* the precondition as we have had to in this book. This benefit is dangerously illusory: there is a proof obligation of implementability in VDM that amounts to exactly the same thing as the calculation of the precondition.

Since there is no recognised structuring mechanism in VDM, we are forced to present the *total* operation all in one go. *Request* has a precondition which is true. Its postcondition must cover both of the possible cases: that there is a block to allocate, and that there is no block left to allocate. *Request* is presented in its entirety in Figure 12.1.

12.3 Algebraic Specification Languages

In Chapter 11 we introduced the algebraic approach to specification. A number of formalisms have been developed which attempt to make this approach practical for specifying large systems by providing mechanisms for structuring their presentation and allowing the re-use of common components. Amongst the better known algebraic

[4] As Jones acknowledges, *postcondition* is a poorly chosen term, since it is describing the relation between the state before and the state after the operation.

$$
\begin{array}{l}
\textit{Request} \quad (u\colon U)\; b\colon B, r\colon \textit{Report} \\
\textsf{ext wr } \textit{dir} \;\; : \textsf{map } B \textsf{ to } U \\
\quad \textsf{wr } \textit{free} \; : \textsf{set of } B \\
\textsf{pre true} \\
\textsf{post} \; (\;\; \overleftarrow{\textit{free}} \neq \{\} \wedge \\
\qquad\quad b \in \overleftarrow{\textit{free}} \wedge \\
\qquad\quad \textit{free} = \overleftarrow{\textit{free}} - \{b\} \wedge \\
\qquad\quad \textit{dir} = \overleftarrow{\textit{dir}} \cup \{b \mapsto u\} \wedge \\
\qquad\quad r = \textit{okay} \;\;) \\
\qquad \vee \\
\qquad (\;\; \overleftarrow{\textit{free}} = \{\} \wedge \\
\qquad\quad r = \textit{fail} \wedge \\
\qquad\quad \overleftarrow{\textit{free}} = \textit{free} \wedge \\
\qquad\quad \overleftarrow{\textit{dir}} = \textit{dir} \;\;)
\end{array}
$$

Figure 12.1: The Request Operation in VDM

specification languages are Clear[5], ACT ONE[6], Larch[7], and variants of OBJ[8].

All of the proposed algebraic specification languages provide a number of *theory building operations* to encourage the writing of structured specifications, and a number of these are described briefly below.

Theory Building Operations

To permit the specification of complex systems by theory presentations we need to establish a *theory of theories* so that algebras of theories can be used to build up large specifications from small components. In what follows we will assume that $T_1, T_2, T_3 \ldots$ denote theory presentations.

Enrichment and Extensions

All algebraic specification languages provide ways in which new theories may be expressed as enrichments or extensions of others. We have used this idea throughout the book, but you should be aware that the terminology is not as standard as we may have implied. Clear adopts a notion of enrichment very similar to the one we have been using, but ACT ONE, for example, would call these *conservative extensions*, saving the name *enrichment* for the special case of a conservative extension where no new sorts are added.

---

[5] Burstall, R.M. and Goguen, J. A., The semantics of Clear, a Specification Language. In *Proceedings of the 1979 Copenhagen Winter School on Abstract Software Specification.*

[6] Ehrig, H. and Mahr, B., Fundamentals of Algebraic Specification 1, Springer Verlag, 1985.

[7] J. V. Guttag, J. J. Horning, and J. M. Wing, Larch in Five Easy Pieces, Digital Systems Research Center Report, July, 1985.

[8] Goguen, J. A. and Tardo, J., An Introduction to OBJ: A Language for Writing and Testing Software Specifications. In *Specification of Reliable Systems*, IEEE, 1979.

Union and Disjoint Union

The simplest form of operator on theories is *union*, which merges two theories, combining the sorts, operations and equations to provide a new theory; any duplicated components are written down only once. The operation is provided by most algebraic specification languages, although it may be given different names; ACT ONE, for example, calls this operation *combine*. We have used unions already, for example in Chapter 11, where we wrote

Data Type: T_3 enrichment of T_1 and T_2

in which the construct T_1 and T_2 formed the union of the two theories prior to the enrichment taking place.

The *union* operation has one major drawback when used for large specifications: every name used in the components becomes global in the union, and hence the writers need to take great care never to use common names where they are not required. This causes problems in achieving modularity, so most specification languages provide a *disjoint union* operation which merges two theories but 'remembers' where the resulting sorts and operations were initially defined, so that any common names can be discriminated by their source. An intuitive way to understand this is to imagine that the *disjoint union* operation prefixes the theory name to each operation and sort before forming the union.

One important difference between *union* and *disjoint union* arises when theories have common subtheories. Consider the following combinations of theories, where *and* denotes union and + denotes disjoint union.

Data Type: T_1 enrichment of T_2
Data Type: T_3 enrichment of T_2
Data Type: T_4 enrichment of T_1 and T_3
Data Type: T_5 enrichment of $T_1 + T_3$

In the case of T_4 the theory clearly contains just one version of the theory T_2, but how many versions are there in T_5? If we extend the disjoint union concept to all subtheories, then we will proliferate copies of T_2. In particular, we might arrive at a situation where there are many theories of *boolean*, so that, for example, there are two terms we intend denoting true, one arising from T_3 and one from T_1, and we will need to explicitly write equations saying that they are equal.

ACT ONE provides ways of expressing both forms of union, + for disjoint union, *and* for union; Clear, however, adopts a different strategy, providing only a disjoint union operation *and*, but performing an ordinary union operation on all subtheories.

Derivation, Renaming and Hiding

To achieve modularity it is sometimes desirable to hide some of the sorts and operations of one theory from another which uses it. In the theory ***high low store***, for example, we used auxillary functions to aid in the process of describing the operations *maxreading* and *minreading*. To achieve reuse of components it is sensible to provide ways in which the terms in a theory can be renamed to reflect their intended use. Most algebraic specification languages provide operations which achieve these ends. OBJ, for example, provides operations for explicitly hiding sorts and operations, and also an *image* mechanism for extracting different *views* of theories. Clear provides a *derive* operation, which allows the required aspects of a theory to be extracted and renamed, ACT ONE provides a *rename* operation allowing sorts and operations to be renamed.

Induce

Clear provides one rather special operation, which allows induction to be used for proofs within a theory. If we view the equations within a theory presentation as providing a rewrite system, then we have a mechanism for showing particular equalities, but not proving generalities. We can show within our theory of sequences, for example, that

$$\#(s \frown t) = \#s + \#t$$

for particular sequences s and t, but we cannot show that

$$\forall s, t \bullet \#(s \frown t) = \#s + \#t$$

The Clear operation *induce* takes a theory and produces a new one within which induction may be used over the structure of the terms. The resulting theory will not, in general, be complete.

Parameterisation and Generics

To allow the widespread reuse of theories, it is necessary to permit generic or parameterised theories. The theory for *sequence*, for example, needs to be parameterised if we want to allow sequences of naturals, sequences of booleans and sequences of reals, without having to write three different theories explicitly. The ways in which this is done differs from language to language. Clear, for example, permits the writing of *theory procedures* which take other theories as arguments, ACT ONE allows parameterised theories which take sorts and operations as actual parameters.

Institutions and Semantics

Underlying an algebraic specification must be a formal system. Moreover, it is usual to have in mind various constraints on the class of models that will be permitted. If our theory of theories is to be reflected by a theory of specifications, then we must capture not only the formal system, but also the constraints on the permitted interpretations. This idea has been formalised into the notion of an *institution*[9]. An institution comprises a *signature* Σ; a set of sentences adhering to the signature, Σ-sentences; a set of Σ-models; and a relation defining what it means for a Σ-model to *satisfy* the Σ-sentences. Equational predicate calculus, with the class interpretations we introduced in Chapter 3 and the notion of initial models we introduced in Chapter 4, constitutes an institution.

ACT ONE, for example, is underpinned by a similar institution. It uses an equational logic, and ensures it is given an initial semantics. This constraint imposes limits on the language's descriptive power similar to those we encountered in Chapter 3. It also prohibits loose

[9] Burstall, R.M. and Goguen, J. A., Introducing Institutions. In *Logics of Programs*, Springer-Verlag, 1983.

specifications, such as a sequence where we wish to leave free the decision as to what the sequence contains. If we wish to give theories expressed in ACT ONE a semantics as mathematical objects, then initial algebras will suffice.

Clear adopts a unique stance, which attempts to overcome these limitations. It is written in such a way that it is parameterised by institution, allowing specifications to be written using any institution. The theory building operations are defined so that they are independent of particular institutions, their semantics leaves them sufficiently free until the institution is identified. To give Clear a mathematical semantics which expresses this parameterisation by institution requires a body of mathematics we have not addressed in this book, called *category theory.*

OBJ is unusual in one major respect, it is executable. This permits the automation and testing of theories. The semantics are initial, but the existence of an interpreter encourages an operational semantics to be given as well. If we wish to view OBJ in this way, there are additional constraints that need to be placed on the logical system. The rewrite rules must *terminate*, thus rules such as

$$add(x,y) = add(y,x)$$

are not allowed, since the interpreter might forever rewrite any expression containing *add*. Also the rules must always rewrite an expression to a unique value, regardless of the order in which they are applied: this property of equations is referred to as the Church-Rosser property.

Larch also adopts an unusual approach, as its logical system is made up of two parts. First there is a *shared language* which allows theories to be written describing objects which are free from implementation details. This is a similar to Clear with an institution of equational logic. There is also a set of *interface languages*, from which one is chosen when an implementation language has been selected. This allow the writing of theories, which are enrichments of those in the shared language, but which reflect the structure of final implementations.

12.4 Process Algebra

The study of concurrency has proved to be a fertile area of research in computing science, yielding many interesting theoretical results. Two important theories of concurrency are *A Calculus of Communicating Systems*[10] and *Communicating Sequential Processes*[11]. Both these theories have produced practical results: the protocol specification technique known as LOTOS[12] is based closely on CCS and ACT ONE; and the programming language for the transputer, occam[13], is based on CSP. Superficially, CCS and CSP look similar: they both describe the behaviour of a concurrent system by describing its *algebra.* However, a deeper examination reveals many differences between the two theories, many of which are very subtle.

A Calculus of Communicating Systems

We shall describe CCS, and note some of the differences between it and CSP as we proceed.

The Basic Calculus

In CCS, there is a set of *names*

$$\Delta = \{\alpha, \beta, \gamma, \delta, \ldots\}$$

and a set of *co-names*

$$\overline{\Delta} = \{\overline{\alpha}, \overline{\beta}, \overline{\gamma}, \overline{\delta}, \ldots\}$$

A name α and a name $\overline{\alpha}$ are called *complementary* names.

Labels are the names or co-names of *actions* which a computing *agent* may perform; they label an agent's *ports.* We call the set of labels Λ, where

---

[10] R. Milner, *A Calculus of Communicating Systems,* Lecture Notes in Computer Science, vol **92**, Springer-Verlag, 1980.

[11] C.A.R. Hoare, *Communicating Sequential Processes,* Prentice-Hall International, 1985.

[12] ISO, Information Processing Systems, Definition of the Temporal Ordering Specification Language, TC97/16 N1987, May, 1984.

[13] INMOS Ltd, *occam Programming Manual,* Prentice-Hall International, 1984.

$$\Lambda = \Delta \cup \overline{\Delta}$$

The *sort* of an agent is simply the set of its labels, and we can regard it as giving information about the agent which is analogous to its *type*. In CSP, a *process* may engage in *actions* which are in its *alphabet*. There is a difference between CCS and CSP here, in that the notion of alphabet is much more powerful than the notion of sort: choosing the alphabet is a vital part of specifying the process's behaviour as well as its "shape".

There is a distinguished event in CCS called τ which is invisible to an observer, and so is not a label. Although no one can detect τ, it plays an important rôle in the theory.

CCS agents are constructed from a small number of combinators, several of which possess quite complicated semantics.

Inaction

The agent that does nothing is denoted *NIL*; it represents *deadlock*.

Action

The agent that first performs an α action and then behaves like the agent Q is denoted

$$\alpha.Q$$

If we give a name to this behaviour, say P, then we write

$$P \Leftarrow \alpha.Q$$

We can describe larger behaviours using recursion; for example the agent

$$R \Leftarrow \alpha.\beta.R$$

can perform an α followed by a β followed by an α followed by a β and so on.

Summation

Choice between different behaviours can be expressed using the summation combinator. The agent

$$R \Leftarrow (\alpha.R + \beta.R)$$

offers a choice between α and β on every step of its recursion. In general, the agent $P + Q$ behaves either like P or like Q. Summation can involve nondeterminism, as is the case in the behaviour expression

$$\alpha.(\beta.NIL + \beta.\gamma.NIL)$$

After the initial α, the agent either behaves like $\beta.NIL$ or like $\beta.\gamma.NIL$, and the choice between these two behaviours is not determined.

In CSP, the issues of choice and nondeterminism are teased apart, and there are separate operators to express each concept.

Composition

The calculus so far allows us to describe the behaviour of just a single agent; composition allows us to describe *communication* between agents. Communication is the *simultaneous* occurrence of *two* complementary actions. We can regard complementary ports as being linked. If agent R contains the port π and agent S contains the port $\overline{\pi}$, then, when a communication occurs over the $\pi/\overline{\pi}$ link, one of three things might happen

1. π may be observed from outside;

2. $\overline{\pi}$ may be observed from outside;

3. R and S may synchronise through the $\pi/\overline{\pi}$ link; this results in an invisible τ action.

The first two possibilities correspond to "interference" by an external agent, the third corresponds to synchronisation of behaviours. Thus, we can say that observation isn't passive: we observe by communicating. Synchronisations can be interfered with by the environment. For example, if we have the behaviours

$$R \Leftarrow \pi.\rho_1.\rho_2.\phi.R$$
$$S \Leftarrow \overline{\pi}.\overline{\phi}.S$$

then we can form the composition of R with S as follows

$$\begin{aligned}
R \mid S &= (\pi.\rho_1.\rho_2.\phi.R) \mid (\overline{\pi}.\overline{\phi}.S) \\
&= \pi.((\rho_1.\rho_2.\phi.R) \mid (\overline{\pi}.\overline{\phi}.S)) \\
&\quad +\overline{\pi}.((\pi.\rho_1.\rho_2.\phi.R) \mid (\overline{\phi}.S)) \\
&\quad +\tau.((\rho_1.\rho_2.\phi.R) \mid (\overline{\phi}.S))
\end{aligned}$$

We have expanded the behaviour of $R \mid S$ according to the rules that we gave: the first line of the expansion shows that π may be observed; the second that $\overline{\pi}$ may be observed; and the third that a successful synchronisation may occur. The first two lines result from interference by observation, and in the third there is the "evidence" of some internal communication.

In CCS the composition operator contains elements of concurrency, nondeterminism, and hiding.

Restriction

A way of avoiding interference with synchronisations is to make their channels internal. The process of doing this abstraction is called restriction. For example, let

$$P \Leftarrow \alpha.\beta.P$$

then we can form a behaviour by restricting the action β or $\overline{\beta}$ as follows

$$\begin{aligned}
P \setminus \{\beta\} &= (\alpha.\beta.P) \setminus \{\beta\} \\
&= \alpha.(\beta.P) \setminus \{\beta\} \\
&= \alpha.NIL
\end{aligned}$$

Since the action α was not being hidden (restricted), it is still available for observation or communication; however, the action β can never be seen externally, and so it can never happen. Thus, the agent is deadlocked after the α action.

It is interesting to note that the concealment operator in CSP, which corresponds to the restriction combinator in CCS, would yield a *different* behaviour from the one that we have just derived. In CSP we would have obtained a process that can engage in the event α repeatedly.

We can give a more complicated example involving communication and restriction

$$P \Leftarrow \alpha.\gamma.P$$
$$Q \Leftarrow \overline{\gamma}.\beta.Q$$

$$\begin{aligned}
(P \mid Q) \setminus \{\gamma\} &= ((\alpha.\gamma.P) \mid (\overline{\gamma}.\beta.Q)) \setminus \{\gamma\} \\
&= \alpha.((\gamma.P) \mid (\overline{\gamma}.\beta.Q)) \setminus \{\gamma\} \\
&= \alpha.\tau.((\alpha.\gamma.P) \mid (\beta.Q)) \setminus \{\gamma\} \\
&= \alpha.\tau.(\alpha.((\gamma.P) \mid (\beta.Q)) \setminus \{\gamma\} \\
&\quad +\beta.((\alpha.\gamma.P) \mid (\overline{\gamma}.\beta.Q)) \setminus \{\gamma\}) \\
&= \alpha.\tau.(\alpha.\beta.\tau.(P \mid Q) \setminus \{\gamma\} \\
&\quad +\beta.\alpha.\tau.(P \mid Q) \setminus \{\gamma\})
\end{aligned}$$

Bisimulation

In CCS two agents are equivalent if their algebraic descriptions are "similar" enough. This notion of similarity is quite precise: if each agent can simulate the other in the sense that we shall give below, then we say that they are ***bisimilar***. In order to understand this notion of bisimulation, define the relation

$$P \xrightarrow{\mu} P'$$

to mean that agent P can perform a μ action, and then behave like agent P'. We can immediately generalise this notion to sequences of actions. If

$$s = \langle \lambda_1, \lambda_2, \ldots, \lambda_n \rangle$$

and

$$P \xrightarrow{\tau} \star \xrightarrow{\lambda_1} \star \xrightarrow{\tau} \star \cdots \xrightarrow{\lambda_n} \star \xrightarrow{\tau} P'$$

then we write

$$P \xRightarrow{s} P'$$

This means that the agent P can perform all the actions in s, interspersed with invisible actions where necessary, and then behave like the agent P'. For example, given

$$P \Leftarrow \tau.\alpha.\beta.\tau.\gamma.P'$$
$$s = \langle \alpha, \beta, \gamma \rangle$$

we have

$$P \stackrel{s}{\Longrightarrow} P'$$

since

$$\begin{aligned} P &\stackrel{\tau}{\longrightarrow} (\tau.\alpha.\beta.\tau.\gamma.P') \\ &\stackrel{\alpha}{\longrightarrow} (\beta.\tau.\gamma.P') \\ &\stackrel{\beta}{\longrightarrow} (\tau.\gamma.P') \\ &\stackrel{\tau}{\longrightarrow} (\gamma.P') \\ &\stackrel{\gamma}{\longrightarrow} P' \end{aligned}$$

We shall denote the set of all process behaviours by $\mathcal{P}$. A relation

$$\_\mathcal{R}\_ : \mathcal{P} \leftrightarrow \mathcal{P}$$

is a bisimulation if, whenever $P\ \mathcal{R}\ Q$, then

$$\begin{array}{lll} \forall\, s : seq[A] \bullet & & \\ (P \stackrel{s}{\Longrightarrow} P' & \Rightarrow & \exists\, Q' : \mathcal{P} \bullet (Q \stackrel{s}{\Longrightarrow} Q') \wedge P'\ \mathcal{R}\ Q') \wedge \\ (Q \stackrel{s}{\Longrightarrow} Q' & \Rightarrow & \exists\, P' : \mathcal{P} \bullet (P \stackrel{s}{\Longrightarrow} P') \wedge P'\ \mathcal{R}\ Q') \end{array}$$

Exercise 12.1 *The proof technique for showing that two processes are equivalent is to exhibit a bisimulation. Given the agents*

$$\begin{aligned} P &\Leftarrow (\alpha.(\beta.NIL + \tau.\gamma.NIL) + \alpha.\gamma.NIL) \\ Q &\Leftarrow \alpha.(\beta.NIL + \tau.\gamma.NIL) \end{aligned}$$

Prove that

$$R = (Id_{\mathcal{P}}) \cup \{P \mapsto Q\}$$

is a bisimulation.

In CSP a different approach is taken to identifying and distinguishing processes. A model of nondeterministic processes is given in which a process is characterised by its alphabet and set of *failures*. A failure is a trace (sequence) of events in the life of the process, and the set of events in which the process can refuse to participate at the end of the trace. This model is strong enough to capture the semantics of nondeterminism, and may be strengthened to handle divergence and real-time behaviours. However, it is different from bisimulation. Define the two process behaviours

$$\begin{aligned} P &\Leftarrow (\alpha.\beta.NIL + \alpha.\gamma.NIL) \\ Q &\Leftarrow (\tau.\alpha.\beta.NIL + \tau.\alpha.\gamma.NIL) \end{aligned}$$

In CSP P and Q would describe the same behaviour, though of course the syntax would be different!

Exercise 12.2 *Prove that P and Q are* not *bisimilar.*

Specifications

The approach to specification in CCS is to construct a member of the equivalence class of behaviour that the specifier wants. Anything that can be shown to be bisimilar to this abstract program is acceptable. For example, suppose that we wanted to describe the operation of a system in which there is a reader and a writer who have mutually exclusive access to some shared data. Our system, T can nondeterministically behave either as a reader, in which it performs a ρ_1 action followed by a ρ_2 action, or as a writer, in which it performs an ω_1 action followed by an ω_2 action. This nondeterminism can be described using an invisible transition to either behaviour

$$T \Leftarrow (\tau.\rho_1.\rho_2.T + \tau.\omega_1.\omega_2.T)$$

A well-known implementation for such a system is to use a semaphore

$$R \Leftarrow \pi.\rho_1.\rho_2.\phi.R$$
$$W \Leftarrow \pi.\omega_1.\omega_2.\phi.W$$
$$S \Leftarrow \overline{\pi}.\overline{\phi}.S$$

Our system is implemented by composing the reader, writer, and semaphore, and then hiding the semaphore's channels

$$I = (R \mid W \mid S) \setminus \{\pi, \phi\}$$

Exercise 12.3 *Prove that T and I are bisimilar*

The approach to specification in CSP is different: a specification is given in the model for the process algebra. Thus, we write predicates that constrain the failures of the process that we wish to implement. We then write a program in the process algebra and prove that it implements the specification using the laws of CSP. The difference between the two techniques is really quite striking.

12.5 Modal Logics

In Chapters 2 and 3 we encountered two formal systems which discussed the notions of truth and falsity, and hence were labelled "logics". Throughout the centuries, philosophers have questioned laws such as those of the excluded middle, contradiction and truth functionality, and have proposed logics which have different sets of laws. They have also questioned issues such as the adequacy of labelling a statement "true" without endorsing this truth as "true by necessity" or "true by possibility". The contention being, for example, that there is a fundamental difference between statements which are true now, and always will be true, and those which just happen to be true at present. Many of these logics were originally developed as part of the continual philosophical debate, but some of the more straightforward of these have since been suggested as useful institutions for writing specifications.

The most developed of these are in the area of *modal logics*, which seek to capture the different notions of truth, such as truth by necessity, via additional operators in the formal system. Typically, a sentence of the form

$$\Box P$$

is used to denote the statement that P is true of necessity, and

$$\Diamond P$$

denotes that P may possibly be true. The traditional interpretation of sentences such as $\Box P$ and $\Diamond P$ introduce the idea of various *worlds*, with necessity being thought of as P being true in all worlds, and possibility meaning that there is at least one world within which P is true.

A particular use of modal logics is the representation of statements which vary over time. By representing time as an ordering of worlds, which can be achieved in a number of different ways, we can use our modalities of truth to indicate statements such as P is always true and P is sometimes true. For example, if we consider time to be discrete, and non-branching, (that is, there is only one past and one future, even though we do not know what it is to be), then we can consider ourselves to be present in the current world, W_0 say. If we ignore the past, then in order to say that P will be true at some stage in the future we can say

$$\Diamond P$$

and if we want to say that P will always be true, we can write

$$\Box P$$

Sometimes it is convenient to introduce the idea that P is true in the next world, this is denoted by

$$\circ P$$

It is important to realise that all of these ideas can be expressed in predicate logic, by explicitly introducing the notion of time into predicates. To say that some object, m, has a particular property at all times, we can introduce a predicate $P(x, t)$ which denotes P holding for object x at time t, and write

$$\forall t\colon \mathit{Time} \bullet P(m, t)$$

Handling time in this way, however, is usually too complex for any but the simplest of systems. Reasoning about time expressed in this way amounts to considering, explicitly, all possible instances of time at which events of interest may occur.

A simple example of the use of this sort of temporal logic is describing components of the states of a sequence of computations. For example, if a computation makes use of two variables, m and n, then we can make statements such as

$$(\Box(m = 7) \land \Diamond(n = 0)) \Rightarrow \Diamond(m + n = 7)$$

These ideas have been captured within a programming language, *Tempura*[14].

One of the more complex uses to which temporal logics have been put is the specification of communication protocols. If we consider the inputs to a communications channel to be recorded in a sequence *inputs*, and the outputs to be recorded in the sequence *outputs*, then the statement

[14] Executing Temporal Logic Programs, Ben Moszkowski, Cambridge University Press, 1986.

$$m \in inputs \Rightarrow \Diamond m \in outputs$$

says that if m is put into the channel it will eventually be sent out of the channel. A more realistic channel might accept the possibility of data being lost, and guarantee to deliver the message only if it is sent more than n times.

$$\#(inputs \restriction \{m\}) > n \Rightarrow \Diamond m \in outputs$$

A variant on this is to consider properties and relations to hold over *intervals* of time, rather than in discrete worlds, and to develop *interval logics* to formalise this idea.

12.6 In Conclusion

In this book, we have concentrated on teaching *principles*: we have described the kind of mathematics that we believe is *essential* to software engineering. In this, we have emphasised two important activities: developing application-oriented theories, and structuring mathematical descriptions. Developing a judicious theory can simplify a specification and proofs of its properties. Exposing structure can help to achieve a simplified description.

These activities are relevant to all the formal methods that we have mentioned in this chapter. Instead of being adherents of Z or of VDM, of CCS or of CSP, of the model-oriented approach or of the algebraic approach, we should be the users of mathematics for software engineering. In the spirit of scientific enquiry, we should seek to understand and become proficient in different formal methods, so that we may see their unifying principles, and know when it is appropriate to apply them[15]. This book has laid the foundations, building the super-structure is up to you.

[15] For a description of the rôles of different specification techniques see B. Cohen, W.T. Harwood & M.I. Jackson, *The Specification of Complex Systems*, Addison-Wesley, 1986.

Index

Glossary

[$\{\}$] empty set (p. 74)
[$\#$] cardinality (p. 144)
[$\Rightarrow$] implication (p. 23)
[$\wedge$] conjunction (p. 22)
[$\{\ldots\}$] set braces (p. 73)
[$\{x : S \mid P(x)\}$] set comprehension (p. 76)
[$\{x : S \mid P(x) \bullet t(x)\}$] set specification in comprehension (p. 79)
[$\{x : S \bullet t(x)\}$] set comprehension by form (p. 77)
[$\bigcap$] distributed intersection (p. 129)
[$\bigcup$] distributed union (p. 128)
[$\rightarrowtail\!\!\!\twoheadrightarrow$] total bijection (p. 141)
[$\cap$] set intersection (p. 126)
[$\frown$] sequence catenation (p. 161)
[$\frown/$] distributed catenation (p. 172)
[$\cup$] set union (p. 125)
[$\dashv$] syntactic turnstile (p. 14)
[$\Leftrightarrow$] double implication (p. 24)
[dom] domain (p. 99)
[$\lhd$] domain restriction (p. 131)
[$\varnothing$] empty set (p. 74)
[$\equiv$] equivalence (p. 26)
[$\exists$] existential quantification (p. 46)
[$\exists!$] unique existential quantification (p. 49)
[$\exists\, x : X \bullet P(x)$] typed existential quantification (p. 84)
[$\exists_1$] unique existential quantification (p. 49)
[$\twoheadrightarrow\!\!\!\!\!\!+$] finite function (p. 145)
[$\rightarrowtail\!\!\!\!\!\!+\!\!+$] finite injection (p. 145)
[$\mathbb{F}$] finite power set (p. 144)
[$\forall$] universal quantification (p. 46)
[$\forall\, x : X \bullet P(x)$] typed universal quantification (p. 84)
[$\rightarrow$] total function (p. 121)
[$=$] equality (p. 57)
[$head$] head of a sequence (p. 166)
[Id_S] identity relation (p. 97)
[$\in$] belongs to (p. 74)
[$\rightarrowtail$] total injection (p. 140)
[$injseq$] set of injective sequences (p. 172)
[$\lambda\, x : S \mid P(x) \bullet t(x)$] lambda expression (p. 122)
[$\leadsto$] rewrites to (p. 228)
[$\vee$] disjunction (p. 23)
[$\mapsto$] maplet (p. 92)
[$\models$] semantic turnstile (p. 25)
[$\widehat{=}$] is equal to by definition (p. 73)
[$\mathbb{N}$] natural numbers (p. 75)
[$\lessdot$] domain subtraction (p. 133)
[$\neg$] negation (p. 22)
[$\notin$] does not belong to (p. 74)
[$\rhd$] range subtraction (p. 134)
[$\oplus$] overriding (p. 137)
[$partition$] partitions of a set (p. 173)
[$\rightarrowtail\!\!\!\!\!\!+\!\!\!\twoheadrightarrow$] partial bijection (p. 141)
[$perm$] set of permutations (p. 172)
[$\nrightarrow$] partial function (p. 118)
[$\rightarrowtail\!\!\!\!\!\!+$] partial injection (p. 138)

[**P**] power set (p. 80)
[$\upharpoonright$] sequence restriction (p. 170)
[$\twoheadrightarrow$] partial surjection (p. 140)
[R^{-1}] inverse (p. 97)
[*ran*] range (p. 99)
[**R**] real numbers (p. 83)
[R^r] reflexive closure (p. 105)
[$\leftrightarrow$] relation (p. 89)
[*rev*] reversal of a sequence (p. 167)
[$\rhd$] range restriction (p. 132)
[R^*] reflexive transitive closure (p. 105)
[$\fatsemi$] relational composition (p. 101)
[$seq[X]$] set of sequences (p. 160)
[$\langle \ldots \rangle$] sequence brackets (p. 161)
[$\setminus$] set difference (p. 127)
[*squash*] squashing of a function into a sequence (p. 169)
[$\subset$] proper subset relation (p. 98)
[$\subseteq$] subset relation (p. 98)
[*succ*] successor function (p. 142)
[$\twoheadrightarrow$] total surjection (p. 140)
[R^s] symmetric closure (p. 105)
[*tail*] tail of a sequence (p. 166)
[$\times$] cartesian product (p. 82)
[R^t] transitive closure (p. 105)
[$..$] subrange (p. 142)
[$\vdash$] syntactic turnstile (p. 13)
[(x, y)] ordered pair of x and y (p. 81)
[$\Box$] necessity operator (p. 279)
[$\Diamond$] possibility operator (p. 279)
[$\circ$] next operator (p. 280)